T0185713

SpringerBriefs in Applied Sciences and Technology

SpringerBriefs present concise summaries of cutting-edge research and practical applications across a wide spectrum of fields. Featuring compact volumes of 50 to 125 pages, the series covers a range of content from professional to academic.

Typical publications can be:

- A timely report of state-of-the art methods
- An introduction to or a manual for the application of mathematical or computer techniques
- A bridge between new research results, as published in journal articles
- A snapshot of a hot or emerging topic
- An in-depth case study
- A presentation of core concepts that students must understand in order to make independent contributions

SpringerBriefs are characterized by fast, global electronic dissemination, standard publishing contracts, standardized manuscript preparation and formatting guidelines, and expedited production schedules.

On the one hand, **SpringerBriefs in Applied Sciences and Technology** are devoted to the publication of fundamentals and applications within the different classical engineering disciplines as well as in interdisciplinary fields that recently emerged between these areas. On the other hand, as the boundary separating fundamental research and applied technology is more and more dissolving, this series is particularly open to trans-disciplinary topics between fundamental science and engineering.

Indexed by EI-Compendex, SCOPUS and Springerlink.

More information about this series at http://www.springer.com/series/8884

Wen Yu · Satyam Paul

Active Control of Bidirectional Structural Vibration

 Springer

Wen Yu
Department of Automatic Control
CINVESTAV - Instituto Politécnico
Nacional
Mexico City, Mexico

Satyam Paul (iD)
Department of Engineering
Design and Mathematics
University of the West of England
Bristol, UK

ISSN 2191-530X   ISSN 2191-5318   (electronic)
SpringerBriefs in Applied Sciences and Technology
ISBN 978-3-030-46649-7   ISBN 978-3-030-46650-3   (eBook)
https://doi.org/10.1007/978-3-030-46650-3

MATLAB and Simulink are registered trademarks of The MathWorks, Inc. See https://www.mathworks.com/trademarks for a list of additional trademarks.

This Springer imprint is published by the registered company Springer Nature Switzerland AG
The registered company address is: Gewerbestrasse 11, 6330 Cham, Switzerland

Contents

1 Active Structure Control 1
 1.1 Introduction 1
 1.2 Structural Control Devices 3
 1.3 Structure Control Algorithms.......................... 9
 1.4 Conclusion.. 14
 References 14

2 Structure Models in Bidirection 19
 2.1 Bidirectional Excitation.............................. 19
 2.2 Structure Model of a Two-Floor Building 24
 2.3 Nonlinear Stiffness 27
 2.4 Conclusion.. 28
 References 28

3 Bidirectional PD/PID Control of Structures 31
 3.1 Introduction 31
 3.2 Active Control of Structural Vibration 32
 3.3 PD/PID Controller of Building Structures 35
 3.4 Stability Analysis 37
 3.5 Experimental Results 47
 3.6 Conclusion.. 54
 References 55

4 Type-2 Fuzzy PD/PID Control of Structures.................. 57
 4.1 Introduction 57
 4.2 PD Control with Type-2 Fuzzy Compensation 58
 4.3 PID Control with Type-2 Fuzzy Compensation 64
 4.4 Experimental Results 70
 4.5 Conclusion.. 77
 References 77

5 Discrete-Time Fuzzy Sliding-Mode Control 79
   5.1 Introduction 79
   5.2 Discrete-Time Model of Building Structure 81
   5.3 Fuzzy Modeling of Structure 82
   5.4 Sliding-Mode Control 87
   5.5 Experimental Results 89
   5.6 Conclusions 95
   References 95

6 Bidirectional Active Control with Vertical Effect 97
   6.1 Introduction 97
   6.2 Tridimensional Model of Structures 99
   6.3 Discrete-Time Model 103
   6.4 Tridimensional Active Control 105
   6.5 Stability Analysis of Tridimensional Active Control 107
   6.6 Experimental Results 110
   6.7 Conclusions 117
   References 117

7 Conclusions 119

Chapter 1
Active Structure Control

1.1 Introduction

Previous studies related to earthquakes like those in 1985 Mexico City, 1994 Northridge, 1995 Kobe, 1999 Kocaeli, 2001 Bhuj, 2008 Sichuan, 2008 Chile, and 2012 Emilia reveal that earthquakes have caused severe damage in civil structures all over the world. The process of modification or to control the building structures from severe damages has become a salient topic in structural engineering. The control of building structures from the hazardous earthquake waves is an area of great interest for the researchers that is growing rapidly [1, 2]. The challenging part of the job lies in the protection of superstructures in the whole of geographic locations from the seismic events, thus providing a means of safer environment for the human occupants. The extensive damages due to an earthquake can be noteworthy and so there is utter necessity to develop effective methods for protection.

The structural control methodology and its applications during earthquakes were first suggested by the researchers more than a century ago. Although Yao in 1972 [3] had proposed the first idea of structural control that played a major role in the advancement of the field of structural engineering, major developments have been noticed during the last 25 years where the structures with preventive systems have been developed. In the area of structural design and its control, the following points should be taken care of:

- The pattern in which the ground and earthquake vibrate during earthquake.
- The design techniques of buildings to withstand earthquakes.
- Innovative strategies for the response control of building structures.

Passive and active control systems play an important role in the response reduction of civil engineering structures subjected to strong seismic vibrations. Passive, active, and semi-active control systems are the most important classes of structural engineering. The two techniques that can be utilized for the control of structural vibrations are as follows:

W. Yu and S. Paul, *Active Control of Bidirectional Structural Vibration*,
SpringerBriefs in Applied Sciences and Technology,
https://doi.org/10.1007/978-3-030-46650-3_1

- Implementation of smart materials in the construction of buildings [4].
- The use of control devices like actuators, dampers, and isolators in the building structures [5].

A worldwide popularity and high demand of structural control and its application had given rise to various researches leading to the publication of many textbooks, for example, [6]. Housner et al. [4] had suggested different types of passive, active, semi-active, and hybrid control systems in his review paper that opens up the importance of control theory in the vibration control of structures. Fisco and Adeli [1] had focused on in-depth studies about active, semi-active, and hybrid control devices along with some control strategies. The main factors affecting the performance of structural control can be categorized as follows:

- Excitation criteria (e.g., unidirectional or bidirectional earthquake and winds).
- Structural characteristics (e.g., natural frequency, degree of freedom and nonlinearity in structures).
- Design of the control system (e.g., device types and quantity, device placements, system models, and control algorithm) [7].

Although most of research has been vested on the seismic analysis considering unidirectional seismic waves, very less researches have been conducted on bidirectional seismic waves. The fact cannot be denied that the earthquake has indeed an arbitrary direction, represented by a bidirectional ground movement [8, 9]. The bidirectional seismic inputs in buildings will induce translation–torsion coupled vibrations in buildings which is more severe with severe structural damage and should be taken into consideration [10]. The intensive research in the field of earthquake engineering revealed the fact that one of the prime factors of building collapse in recent times is asymmetric building structures under the grip of bidirectional seismic ground motions [11].

The active devices are capable of adding forces to the structures. If the control forces for these active devices are generated by unstable controller, then it may cause unusual vibrations to the building structure, thus affecting the performance. So it is utter necessary to analyze the stability of the controller. Also, it is very important to study the controllers' performance and effectivity under the effect of bidirectional forces. The bidirectional forces acting on the building will result in torsion in the building which is an important area of research and should be taken into consideration. Apart from that, the experimental verifications of these controllers in mitigation of bidirectional seismic waves were not given due to consideration. Hence, the implementation of a controller will be challenging if these issues are not handled in an efficient manner. It is an important aspect of structural control system to sense response of the structure continuously and give efficient response in order to mitigate the vibration caused by seismic waves. Several types of controllers were implemented to attenuate the structural vibrations due to unidirectional earthquake. So more focus should be vested in control action involving bidirectional forces. Torsion is an important aspect of buildings under the effect of earthquake and needs to be dealt with in an efficient manner. The designed controller should possess the capability of measuring the response and act on the controller mechanism of the damper and actuator in order to minimize the vibrations. This is done to regulate the output in order to keep the system states such as position and velocity very close to zero. An innovative and efficient control design will broaden the effectiveness of bidirectional vibration mitigation. One more important criterion while designing a controller is its stability. The controller instability will result in the unsuitable system operation and consequently may incur significant damages to the building ultimately causing harm to humans. Another important aspect is parameter uncertainty incorporated in the buildings. So it is essential to design an innovative controller that requires minimum system parameters. Stability and robustness are the important criteria that should be taken into consideration while proposing a high-performance controller. Finally, the performance of the proposed controller should be verified under the impact of bidirectional seismic waves.

Based on the above discussions, the objectives of this book are enlisted as follows:

- A need of developing a novel mechanism to control the lateral–torsional vibration due to bidirectional forces on the structure.
- To propose a high-performance controller design that involves least information on structural parameters.
- To validate the stability of proposed controller from theoretical point of view.
- To verify experimentally the performance of the proposed algorithms and to compare the performance of the controller based on the experimental analysis in order to validate the most superior controller for the mitigation of structural vibration.

1.2   Structural Control Devices

Vibration suppression in an appropriate quantity can prevent the structures from fracture or collapse. Some devices play this suppression role to prevent the structure from damages. The control devices, such as actuators, isolators, and dampers, are installed to suppress the external vibrations. These structural control devices are getting more popularity and attention along with their applications in building structures. The structural control devices for the seismic hazards can be categorized as passive, active, hybrid, and semi-active [12]. In the last two decades, the active, semi-active, and hybrid control are paid more attentions than the passive devices [13]. The conception and characteristic of the structural control devices for bidirectional seismic waves are illustrated sequentially.

A passive control device is incorporated into a structure. It modifies the stiffness or the damping of the structure in a suitable way. The passive control system does not require an external power source for its operation. It generates control force opposite to the motion of controlled structured system [14]. The passive systems can be divided into two basic categories: (1) base isolation systems and (2) energy dissipation systems. There are many passive control devices, for example, viscoelastic dampers, tuned mass dampers, frictional dampers, tuned liquid dampers, and base isolation systems [15]. The principal function of a passive energy dissipation system is to reduce the inelastic energy dissipation demand on the framing system of a structure [16].

The main drawback of the passive control devices is that they cannot adapt to the change of the natural frequency caused by the structural nonlinearity and huge seismic excitations, especially for multiple floor buildings [1], although multiple and tuned dampers can be applied for different frequencies. Since 1970s, remarkable progress has been made in the field of active control of civil engineering structures subjected to natural forces such as winds and earthquakes [13]. The active structure control modifies the structural motion by some external forces. Topics covered on active structural control can be found in [17]. Compared with the passive devices, the active systems have the following advantages [15]:

1. Motion control can be achieved with greater effectivity.
2. In account of ground motions, it is relatively insensitive.
3. It can be applied to the multi-hazard remission circumstances.
4. Control objectives can be selected flexibly.

The forces of the passive control devices solely depend on the structural motion. The tuned mass damper (TMD) is considered to be an energy dissipation system, although the primitive concept of this system is not to dissipate energy. It transfers the energy from the building structure to the tuned mass dampers (TMDs) (absorbers). The basic principle of TMD is to obtain optimal damping parameters, in order to control the displacement of an undamped system subjected to a harmonic force [18]. The coupled lateral–torsional motions under seismic excitations are exhibited by the building structures with intended eccentricities between their mass and stiffness centers. In [19], investigation of TMDs in arrangements termed as coupled tuned mass dampers (CTMDs) was carried out, where translational springs and viscous dampers are used to connect mass in an eccentric manner. The CTMD works in coupled mode that includes lateral and rotational vibrations. This technology is utilized to control coupled lateral and torsional vibrations of asymmetric buildings. The results revealed that CTMDs are more effective and robust in controlling coupled lateral and torsional vibrations of asymmetric buildings.

In [20], multiple tuned mass dampers (MTMDs) were proposed with distributed natural frequencies. Several researches had been carried out to establish the effectiveness of MTMDs, and it had been verified that MTMDs had advantages over single TMD. A multiple tuned mass damper (MTMD) system is shown in Fig. 1.1. It consists of the main system, which has n TMDS with different dynamic characteristics. The main system is subjected to a lateral force. The main system and each TMD vibrate in the lateral direction. Due to torsional coupling, the main system has torsional vibration. A state-of-the-art review on the topic of response control of structures utilizing the passive tuned mass damper(s) is illustrated in [21]. Tuned liquid column damper (TLCD) has uniform cross section with U-shaped tube attached.

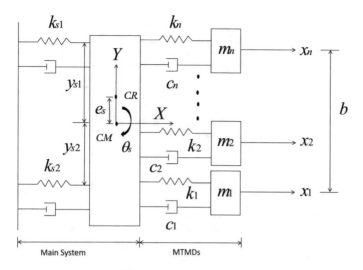

Fig. 1.1 The multiple tuned mass damper (MTMD)

Fig. 1.2 The tuned liquid column damper (TLCD)

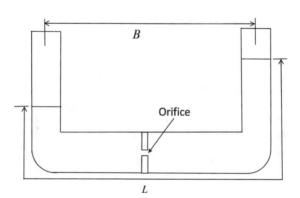

The schematic view has been shown in Fig. 1.2. The vibrational energy from the structure is transferred to the TLCD liquid via the movement of the rigid TLCD container, thus stimulating the TLCD liquid.

In [22], the methodology of vibration control of eccentric structures using TLCD modeled as torsionally coupled multi-storey shear structures which is under the grip of multidimensional seismic excitations has been investigated.

The circular tuned liquid column damper (CTLCD) is shown in Fig. 1.3. This advance control device is highly responsive to the torsion. CTLCD can be applied for both torsional vibration and torsionally coupled vibration. The effectiveness of CTLCD for the structural torsional response is studied by [23]. Stochastic vibration theory is applied to identify the optimal parameters of CTLCD in [24].

In [9], a new type of control device termed as tuned liquid mass damper (TLMD) was presented in order to control the torsional response of building structures

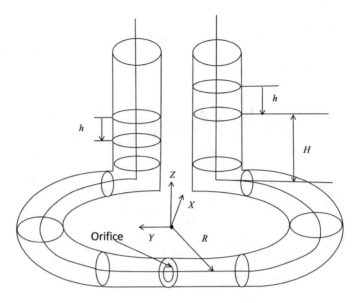

Fig. 1.3 The circular tuned mass damper (CTLCD)

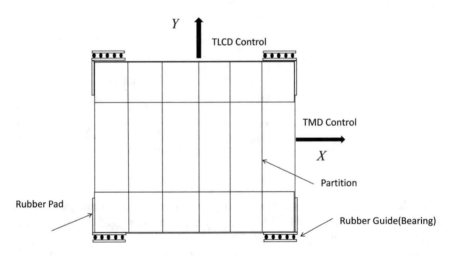

Fig. 1.4 The tuned liquid mass damper (TLMD)

subjected to bidirectional earthquake waves. The mass of TLMD includes both TLCD tank and the liquid in the tank. The stiffness is compensated by natural rubbers. The main working concept of TLMD is to operate a TLCD in one direction, and run a TMD in the other orthogonal direction, see Fig. 1.4. In [25], the control performance of the novel sealed, torsional tuned liquid column gas damper (TTLCGD) in order to minimize the coupled flexural torsional response of plan-asymmetric buildings under the grip of seismic loads has been discussed. The analysis of technique associated reveals that TTLCGD is an effective control device in suppressing the time-harmonic excitation and the earthquake response.

In [26], a new performance index for active vibration control of three-dimensional structures was proposed. To analytically prove the existence of the proposed performance index, a six-storey three-dimensional structure is taken into consideration as an example with a fully active tendon controller system implemented in one direction of the building. The building under analysis is modeled as a structure made up of members joint by a rigid floor diaphragm in a manner so that it has three degrees of freedom at each floor, lateral displacements in two perpendicular directions, and a rotation with respect to a vertical axis for the third dimension. An optimal design of active tuned mass damper for minimizing the vibration of irregular buildings along translational–torsional direction was investigated in [27].

Semi-active control devices are regarded as controllable passive devices. The main objective of these devices is saving control resources. The actuators of the semi-active control do not add mechanical energy to the structure directly. The power break down semi-active control system offers some degrees of protection with the help of embedded passive components. The semi-active devices take the advantages of the passive and the active controls. It requires less power than the active control devices. They can even be operated by the battery in the case of power failure during the seismic event [13]. They perform significantly better than passive devices. An exhaustive review on the semi-active devices is proposed in [12]. The magnetorheological (MR) damper is the most popular semi-active damping device. It works on the magnetorheological fluid and is controlled by a magnetic field. Generally, the magnetic field is produced by electromagnet. It requires minimal power for its operation. The suspended minute iron particles in a base fluid are termed as MR fluids. This type of liquids has the capability of changing from free-flowing linear viscous state to semi-solid state with controllable yield strength under a magnetic field. The result of uncovering the liquid to a magnetic field is the particles' use the form of chains. These chains obstruct the flow and solidify the fluid in a span of milliseconds. The stress is directly proportional to the magnitude of the applied magnetic field [28]. The behavior of MR fluid can be simulated by the Bingham plastic model, which is an extension of the Newtonian flow. The other way of determining the behavior of MR fluid is to analyze the yield stress of the fluid. The MR damper contributes significantly to the field of civil engineering. In [29], a prototype shear-mode MR damper is proposed.

Hybrid base isolation (HBI) had been a matter of interest for a number of researchers due to its effectivity and consists of a passive base isolation system combined with a control actuator to generate the effects of the base isolation system. Several researches on base isolation system have been carried out and installed in several structural engineering projects due to its positive attributes like simplicity, reliability, and effectiveness (Fig. 1.5).

In [30], the application of hybrid mass damper (HMD) system consisting of TMD and active mass damper (AMD) to control torsionally coupled building structures under bidirectional seismic force was proposed. In this context, the fuzzy

Fig. 1.5 Hybrid mass
damper (HMD) system
installed in Nth floor

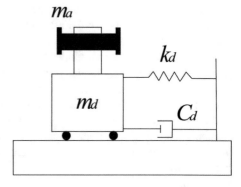

Fig. 1.6 Hybrid mass
damper (HMD) system
installed in Nth floor

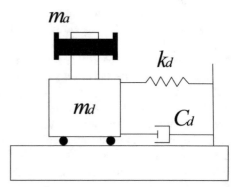

logic controller is used to control the HMD system. Complex structural systems
have nonlinearities and uncertainties in both the structural properties and the magnitude of the loading. Thus, it is difficult to derive and identify an appropriate and
accurate dynamic model for designing the traditional controller. An intelligent controller can be designed without specifying a very precise and accurate dynamic model
of the structure. Such an intelligent controller has been introduced, using a fuzzy logic
control system. The schematic view is shown in Fig. 1.6.

Kim and Adeli [31] had investigated hybrid damper-TLCD control system to control 3D coupled irregular buildings subjected to bidirectional seismic waves. Simulation results for control of two multi-storey moment-resisting space steel structures
with vertical and plan irregularities show clearly that the hybrid damper-TLCD control system significantly reduces responses of irregular buildings subjected to various
earthquake ground motions as well as increases reliability and maximum operabil-
ity during power failure. The comparisons between uncontrolled, passive control,
active control, semi-active control, and hybrid control devices are demonstrated in
Fig. 1.7 [13].

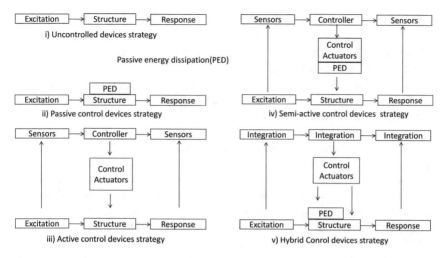

Fig. 1.7 Comparison of control devices

1.3 Structure Control Algorithms

The appropriate design of a controller is utter necessary so that it can send essential control signal to the control devices in order to reduce the structural responses. The main strategy involved within the control scheme is to prevent the collapse of structures under bidirectional seismic waves by controlling the coupled translation–torsion response of the structures [32–35]. In this section, various control strategies on the basis of various techniques are presented. The main objective of the bidirectional control is to change the coupled translation–torsion response of the building structure, in order to prevent the collapse of building under bidirectional seismic waves [36]. Robustness, fault tolerance, simplicity, and realizability criteria are considered [37].

Time delay from the measurement of the actuator is a limit for vibration control. The control loop includes vibration data measurement, data filtration, control algorithm, data transmission, and actuation. The control loop has also phase shift by time delay [38]. The time delay may cause instability in the closed loop [39]. A review of time-delay compensation methods can be found in [40].

The proper placement of sensing and control devices is an important research field of structural control. It results in the effective measurement and control operation. It also affects the controllability and observability of the controlled system [41, 42]. In [43], the location performance index of the actuator and sensor are presented, which can be computed by the Hankel singular values.

In [44], the placing of the sensor at the center of mass is suggested. The proposal validated that the center of mass may not be good for the sensor position. Arbitrary arrangement of sensors is better subjected to bidirectional seismic motion. In [2], a detailed survey of the optimal placement of control devices was presented. In [45], energy dissipation was utilized to analyze the position of the controller, in order to minimize translation–torsion coupling effects. It suggests the locations which are nearby to the geometric center of the structure can minimize the torsional effect.

The working principle of PID controller is based on the feedback error $e(t)$ which is otherwise used to calculate the required control force. In the case of structural applications if the desired state is in the equilibrium position, then the reference signal is considered to be taken as zero. The principle of PID control is to use the feedback error $e(t)$, which is the difference between the output signal $y(t)$ and the reference signal $r(t)$. Once the error is calculated, the main aim of the controller is to minimize the error for the next iteration process by carefully manipulating the inputs.

It is the most popular industrial controller. A comparison between a sliding -mode control and PID control for the structural system is investigated by [46]. In [47], the effects of measure seismic waves on a six-storey asymmetric structural model compiled with frictional dampers were investigated. The methodology deals with the control of torsional response of asymmetric structures and to obtain a lower level of torsional balance by arranging empirical center of balance (ECB) of the structure at the same distance from the edges of the building plan. The axial displacement of each actuator is controlled using a conventional PID controller. In this research, frictional dampers proved its effectiveness of controlling lateral–torsional coupling of torsionally flexible as well as stiff structures. The most important optimal controllers are the linear quadratic regulator (LQR) and linear quadratic Gaussian (LQG) control.

In [48], a semi-active control to the coupled translational and torsional vibration of a two-storey asymmetric building subjected to seismic excitations was presented. An LQG controller is involved as a nominal linear controller, considering the ground acceleration with white noise. In [49], active isolation was implemented and conducted experiments in order to verify the behavior of seismically excited buildings under multidirectional earthquake force. Active isolation technique works in combination with base isolation system and controllable actuators. The base isolation methodology offers effective approach in reducing interstorey drifts and floor accelerations that works in phase with the adaptive nature of the active system in order to generate higher level performance against wide range of earthquakes. In this methodology, LQG control steps are obtained using LQR and Kalman estimator.

In [31], the control of 3D coupled irregular buildings subjected to bidirectional seismic waves was investigated. To find the optimal control forces, a wavelet-based algorithm involving optimal control is utilized. It has been suggested in their work that LQR or LQG algorithm can be used as a control algorithm for the feedback controller as per the investigation mentioned [15, 50, 51]. In [52], a sequential optimal control for serially connected isolated structure subjected to bidirectional earthquake was suggested. Sequential control algorithm has inherent capabilities to construct control objective function under bidirectional earthquake situations.

H^∞ control methodology has been relied on as an effective approach in structural vibration control which is classified as linear robust control. This scheme is unresponsive to the disturbances and parametric differences and so it is most preferred for multiple-input multiple-output (MIMO)-type structural control systems [53]. Design method of H^∞ control system and its effectiveness was presented by [54]. The analysis

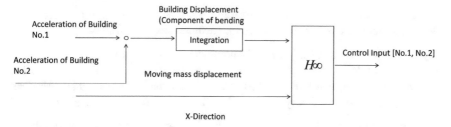

Fig. 1.8 *X*-direction

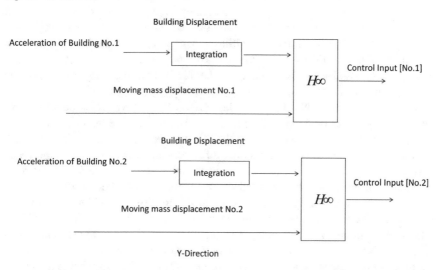

Fig. 1.9 *Y*-direction

was carried out on 23-storey building in Tokyo using a pair of hybrid mass dampers. Bidirectional seismic excitations were considered during the investigation. The control technique was established by taking into consideration *x*- and *y*-directions separately. The bending component of the vibration was controlled along *x*-direction only while along the y-direction, control of bending and torsion is considered. The scheme of the control system is shown in Figs. 1.8 and 1.9.

In [55], the use of a robust optimal H∞control for the two AMD systems was elaborated. The AMD system was placed on the top of the unsymmetrical building for the vibration control. The building was subjected to bidirectional seismic excitations. The H∞control uses the technique of LMI-based solution blended with robustness specifications. H∞ direct output feedback control of buildings under bidirectional acceleration considering the effects of soil–structure interaction was investigated by [56].

Sliding-mode control (SMC) is designed for uncertain nonlinear systems [53]. It is effective in terms of robustness against the changes in the parameters and external disturbances. It has been successfully applied for structural control [57].

In [58], SMC is used to control bending and torsional vibration of a six-storey flexible structure. The controller takes into accounts two conditions:

(1) controller design considering only nonlinear control inputs.
(2) controller design considering nonlinear and sub-equivalent control inputs. The important feature of SMC is robustness under the uncertainties and disturbances. Lyapunov stability theory is implemented to prove the system stability in [57].

A neural network (NN) is characterized by (1) an area which consists of number of neurons along with their interconnections and layers; (2) its technique of implementing the weights on the connections and is termed as learning algorithm. In [59], an NN-based emulator computes the response of a 2D frame structure involving three-storey building. The feedforward multilayer perceptron with the backpropagation algorithm is used in [37] for structure control. [60] presents a wavelet neural network (WNN)-based active nonlinear controller for 3D buildings subjected to seismic excitation in both x- and y-directions.

The combination of NN with the classical control theory yields better control results than conventional controllers [61, 62]. The hybrid intelligent control algorithm applied to semi-active control of the MR damper is presented in [55]. In [63], a direct adaptive neural controller subjected to bidirectional earthquake inputs was presented. Both the system parameters and the nonlinear estimation of force have uncertainties, which can be canceled by the adaptive controller. In [64, 65], NN for the structural reliability analysis was utilized. In [66], an NN-based prediction scheme was proposed for the dynamic behavior of structural systems under multiple seismic excitations. The NN prediction includes two different ways: (1) A non-adaptive scheme that uses multiple accelerometers in training NN, and utilizes for the prediction of the structural seismic response; (2) An adaptive scheme uses multiple accelerometers in the training. The application of RBF neural network for the mitigation of structural vibration is presented in [67].

Linguistic criterion is an effective feature of fuzzy control rules that can be easily modified and understood clearly [68]. In [30], a fuzzy logic controller with multipurpose optimal design was proposed to drive HMD for the response control of the torsionally coupled seismically excited buildings. HMD system consists of four HMDs arranged in such a way that this system can control the torsional mode of vibration effectively in addition to the texture modes of vibration. The design of the fuzzy logic controller (FLC) based on the selection procedure that includes five membership functions for each of the input variables and seven membership functions for the output variable.

The minimization of structural torsion responses using semi-active dampers has been presented by [69]. In their investigation, the MR damper is employed for the real-time control of the response of structures under seismic excitations. The methodology of fuzzy modeling of MR dampers has been shown in [70]. In [71], supervisory fuzzy controller was implemented to control two lower level fuzzy controllers. In [72], it has been illustrated that the dynamic fuzzy wavelet NN can precisely forecast structural displacements. In [60], the wavelet neural network (WNN) model-based active nonlinear controller for the response control of 3D buildings subjected to seismic excitation in both x- and y-directions has been presented. The main aim is to control the torsional and lateral motions of 3D irregular structures. The structural responses are predicted using a dynamic fuzzy WNN which is a fuzzy wavelet neuroemulator. Estimation of future time steps is utter necessary to control the structural responses effectively. This method is essential in determining the magnitude of the required control forces.

In 1975, Holland was the first to propose the general scheme of genetic algorithm (GA). GA uses natural genetic theory to build an optimal search algorithm [73]. A GA can be divided into three parts [74]:

1. Code and decode the variables into the strings form.
2. Evaluate the fitness of each solution string.
3. Evaluate the strings of the next generation by applying genetic operators.

In [75], GA is utilized to MR dampers in the reduction of translation–torsion coupled responses of an asymmetric structure. The experiment was carried out at the State Key Lab of Coastal and Offshore Engineering at Dalian University of Technology. The parameters of the multi-state control strategy (MSC) which utilizes the velocity response as the state-switch parameter are optimized by GA method. This MSC is developed in the intention to control torsional seismic response of an asymmetric structure. In their research, also the threshold vector of the MR damper is optimized using GA. The parameters from the velocity response and the threshold vector of the MR damper are optimized by the GA method. In [76], a new neurogenetic algorithm was presented to evaluate the optimal control forces for active control of 3D building structures. It includes geometrical and material nonlinearities, coupling action between lateral and torsional motions, and actuator dynamics. In this case, a floating-point GA was used. The methodology used can be categorized as follows:

(i) representation of chromosomes,
(ii) initial population,
(iii) function related to fitness,
(iv) selection function,
(v) genetic operator, and
(vi) termination scheme.

The study results suggest that the new control technique efficiently reduces the response of two irregular 3D building structures under seismic inputs including structures with plan and irregular elevation. The study results suggest that the new control technique efficiently reduces the response of two irregular 3D building structures under seismic inputs including structures with plan and irregular elevation.

In [60], a new nonlinear control model for the active control of a 3D building structure was developed. The optimal control forces are computed with the floating-point GA. GA can help to decide the positions of the control devices [77]. The coupled torsional–lateral response is attenuated by a semi-active control under bidirectional seismic input. In [78], technique of reducing the seismic effects of the spatial structures by the installation of MR dampers was proposed. It uses small populations to solve the optimization problem embedded in the semi-active control. GA is used to optimize dampers' passive parameters and controller gain in [79].

The concept of absorber system with multi-objective optimal design for torsionally coupled earthquake excited structures is presented by [30]. It uses a multi-objective version of GA to extract the design parameters of absorber system. The two-branch tournament GA as mentioned by [80] extends two-branch tournament GA to three-branch tournament GA and applies to the multi-objective optimization of the TMD system.

1.4 Conclusion

In this chapter, the structural control techniques of structures subjected to bidirectional earthquake are considered. The main difference with normal structure controllers is the lateral–torsional coupled response. We discuss recent new techniques, methodology, and concepts in these areas. We focus on all important results in last two decades in the field of structural engineering with respect to the bidirectional earthquakes. The important observations from this chapter are as follows:

1. Most of the existing researches only consider the structure control under unidirectional seismic wave. This chapter explores the effects of bidirectional seismic waves, which is normal for the real earthquake.
2. Real buildings are generally asymmetric in nature to some extent. This criterion induces lateral and torsional vibrations in combination.
3. The reduction of translational and torsional response of structures often involves the usage of multiple dampers [81].
4. Little research to sliding-mode control is carried out in order to reduce translation–torsion coupled vibration with bidirectional seismic inputs.
5. In case of building structures subjected to multiple excitations, the use of online identification technique is better.
6. The intelligent control like fuzzy logic is favored for the structural control, because it does not require system information.
7. PD/PID controller is robust, fault-tolerant, and very easy to implement.

References

1. N.R. Fisco, H. Adeli, Smart structures: part I—active and semi-active control. Scientia Iranica **18**(3), 275–284 (2011)
2. N.R. Fisco, H. Adeli, Smart structures: part II—hybrid control systems and control strategies. Scientia Iranica **18**(3), 285–295 (2011)
3. J.T.P. Yao, Concept of structural control. J. Struct. Div. **98**(7), 1567–1574 (1972)

4. G.W. Housner, L.A. Bergman, T.K. Caughey, A.G. Chassiakos, R.O. Claus, S.F. Masri, R.E. Skeleton, T.T. Soong, B.F. Spencer Jr., J.T.P. Yao, Structural control: past, present and future. J. Eng. Mech. **123**(9), 897–971 (1997)
5. R.J. McNamara, Tuned mass dampers for buildings. J. Struct. Div. **103**(9), 1785–1798 (1977)
6. B. Donaldson, *Introduction to Structural Dynamics* (Cambridge University Press, UK, 2006)
7. F. Yi, S.J. Dyke, Structural control systems: performance assessment, in *American Control Conference*, vol. 1, no. 6 (2000), pp. 14–18
8. E. Cruz, S. Cominetti, Three-dimensional buildings subjected to bidirectional earthquakes. Validity of analysis considering unidirectional earthquakes, in *12th World Conference on Earthquake Engineering* (2000)
9. J.S. Heo, S.K. Lee, E. Park, S.H. Lee, K.W. Min, H. Kim, J. Jo, B.H. Cho, Performance test of a tuned liquid mass damper for reducing bidirectional responses of building structures, in *The Structural Design of Tall and Special Buildings,* vol. 18, no. 7, (2009), pp. 789–805
10. J. Zhang, K. Zeng, J. Jiang, An optimal design of bi-directional TMD for three dimensional structure. Comput. Struct. Eng. 935–941 (2009)
11. J.L. Lin, K.C. Tsai, Seismic analysis of two-way asymmetric building systems under bidirectional seismic ground motions. Earthq. Eng. Struct. Dyn. **37**(2), 305–328 (2008)
12. B.F. Spencer, S. Nagarajaiah, State of the art of structural control. J. Struct. Eng. **129**(7), 845–856 (2003)
13. T.T. Soong, B.F. Spencer, Supplemental energy dissipation: state-of-theart and state-of-the-practice. Eng. Struct. **24**(3), 243–259 (2002)
14. S.G. Luca, F. Chira, V.O. Rosca, Passive active and semi-active control systems in civil engineering, Constructil Arhitectura 3–4 (2005)
15. T.T. Soong, *Active Structural Control: Theory and Practice* (Addison-Wesley Pub, New York, 1999)
16. M.C. Constantinou, M.D. Symans, Seismic response of structures with supplemental damping, in *The Structural Design of Tall Buildings,* vol. 22, no. 2 (1993), pp. 77–92
17. T.K. Datta, A state-of-the-art review on active control of structures. ISET J. Earthq. Technol. **40**(1), 1–17 (2003)
18. J.P. Hartog, *Mechanical Vibrations* (McGraw-Hill, New York, 1956)
19. N.B. Desu, S.K. Deb, A. Dutta, Coupled tuned mass dampers for control of coupled vibrations in asymmetric buildings, in *Structural Control and Health Monitoring,* vol. 13, no. 5 (2006), pp. 897–916
20. K. Xu, T. Igusa, Dynamic characteristics of multiple substructures with closely spaced frequencies. Earthq. Eng. Struct. Dyn. **21**(12), 1059–1070 (1992)
21. S. Elias, V. Matsagar, Research developments in vibration control of structures using passive tuned mass dampers. Ann. Rev. Control **44**, 129–156 (2017)
22. H.-N. Li, L.-S. Huo, Seismic response reduction of eccentric structures using tuned liquid column damper (TLCD), in *Vibration Analysis and Control—New Trends and Development* (2011)
23. M.J. Hochrainer, C. Adam, F. Ziegler, Application of tuned liquid column dampers for passive structural control, in *7th International Congress on Sound and Vibration (ICSV 7),* Garmisch-Partenkirchen, Germany (2000)
24. S.G. Liang, Experiment study of torsionally structural vibration control using circular tuned liquid column dampers. Spec Struct. **13**(3), 33–35 (1996)
25. C. Fu, Application of torsional tuned liquid column gas damper for plan-asymmetric buildings, in *Structural Control and Health Monitoring,* vol. 18, no. 5 (2011), pp. 492–509
26. A. Yanik, J.P. Pinelli, H. Gutierrez, Control of a three-dimensional structure with magneto-rheological dampers, in *11th International Conference on Vibration Problems,* ed by Z. Dimitrovová et al., Lisbon, Portugal (2013)
27. M. Azimi, H. Pan, M. Abdeddaim, Z. Lin, Optimal design, of active tuned mass dampers for mitigating translational-torsional motion of irregular buildings, in *Experimental Vibration Analysis for Civil Structures (EVACES),* ed by J. Conte, R. Astroza, G. Benzoni, G. Feltrin, K. Loh, B. Moaveni. Lecture Notes in Civil Engineering, vol. 5 (Springer, Cham, 2017), p. 2018

28. M.R. Jolly, J.W. Bender, J.D. Carlson, Properties and applications of commercial magnetorhe-ological fluids, in *Smart Structures and Materials 1998: Passive Damping and Isolation*, vol. 3327 (1998), pp. 262–275

29. F. Yi, S.J. Dyke, J.M. Caicedo, J.D. Carlsonf, Experimental verification of multi-input seismic control strategies for smart dampers. J. Eng. Mech. **127**(11), 1152–1164 (2001)

30. A.S. Ahlawat, A. Ramaswamy, Multiobjective optimal FLC driven hybrid mass damper system for torsionally coupled, seismically excited structures. Earthq. Eng. Struct. Dyn. **31**(12), 2121–2139 (2002)

31. H. Kim, H. Adeli, Hybrid control of irregular steel highrise building structures under seismic excitations. Int. J. Numer. Methods Eng. **63**(12), 1757–1774 (2005)

32. J.M. Angeles-Cervantes, L. Alvarez-Icaza, 3D Identification of buildings seismically excited, in *16th IFAC World Congress,* vol. 16, Czech Republic (2005)

33. V. Gattulli, M. Lepidi, F. Potenza, Seismic protection of frame structures via semi-active control: modeling and implementation issues. Earthq. Eng. Eng. Vibr. **8**(4), 645–672 (2009)

34. J.L. Lin, K.C. Tsai, Y.J. Yu, Bi-directional coupled tuned mass dampers for the seismic response control of two-way asymmetric-plan buildings. Earthq. Eng. Struct. Dyn. **40**(6), 675–690 (2011)

35. B. Zhao, H. Gao, Torsional vibration control of high-rise building with large local space by using tuned mass damper. Adv. Materi. Res. **446–449**, 3066–3071 (2012)

36. M.P. Singh, S. Singh, L.M. Moreschi, Tuned mass dampers for response control of torsional buildings. Earthq. Eng. Struct. Dyn. **31**(4), 749–769 (2002)

37. Y. Tang, Active control of SDF systems using artificial neural networks. Comput. Struct. **60**(5), 695–703 (1996)

38. R. Alkhatib, M.F. Golnaraghi, Active structural vibration control: a review, in *The Shock and Vibration Digest,* vol. 35, no. 5 (2003), pp. 367–383

39. M.D. Symans, M.C. Constantinou, Semi-active control of earthquake induced vibration, in *World Conference on Earthquake Engineering* (1996)

40. A.K. Agrawal, J.N. Yang, Compensation of time-delay for control of civil engineering struc-tures. J. Earthq. Eng. Struct. Dyn. **29**(1), 37–62 (2000)

41. F. Amini, M.R. Tavassoli, Optimal structural active control force, number and placement of controllers. Eng. Struct. **27**(9), 1306–1316 (2005)

42. O.I. Obe, Optimal actuators placements for the active control of flexible structures. J. Math. Analy. Appl. **105**(1), 12–25 (1985)

43. W. Gawronski, Actuator and sensor placement for structural testing and control. J. Sound Vibr. **208**(1), 101–109 (1997)

44. J.M. Angeles-Cervantes, L. Alvarez-Icaza, 3D Identification of buildings seismically excited, in *16th IFAC World Congress,* vol. 16, Czech Republic (2005)

45. B. Wu, J.P. Ou, T.T. Soong, Optimal placement of energy dissipation devices for three-dimensional structures. Eng. Struct. **19**(2), 113–125 (1997)

46. R. Guclu, Sliding mode and PID control of a structural system against earthquake. Math. Comput. Modell. **44**(1–2), 210–217 (2006)

47. I.J. Vial, J.C. de la Llera, J.L. Almazan, V. Ceballos, Torsional balance of plan-asymmetric structures with frictional dampers: experimental results. Earthq. Eng. Struct. Dyn. **35**(15), 1875–1898 (2006)

48. O. Yoshida, S.J. Dyke, L.M. Giacosa, K.Z. Truman, Experimental verification on torsional response control of asymmetric buildings using MR dampers. Earthq. Eng. Struct. Dyn. **32**(13), 2085–2105 (2003)

49. C.-M. Chang, B.F. Spencer Jr., P. Shi, Multiaxial active isolation for seismic protection of buildings, in *Structural Control and Health Monitoring,* vol. 21 (2014), pp.484–502

50. H. Adeli, A. Saleh, Optimal control of adaptive/smart bridge structures. J. Struct. Eng. **123**(2), 218 –226 (1997)

51. R.E. Christenson, B.F. Spencer Jr., N. Hori, K. Seto, Coupled building control using acceleration feedback, in *Computer-Aided Civil and Infrastructure Engineering,* vol. 18, no. 1 (2003), pp. 4–18

52. Y. Du, Z. Lin, Sequential optimal control for serially connected isolated structures subject to two-directional horizontal earthquake, in *Control and Automation (ICCA)*, Xiamen (2010), pp. 1508–1511
53. V.I. Utkin, *Sliding Modes in Control and Optimization* (Springer, Berlin, 1992)
54. T. Fujinami, Y. Saito, M. Morishita, Y. Koike, K. Tanida, A hybrid mass damper system controlled by H∞ control theory for reducing bending—torsion vibration of an actual building. Earthq. Eng. Struct. Dyn. **30**(11), 1639–1653 (2001)
55. Z. Li, S. Wang, Robust optimal H∞ control for irregular buildings with AMD via LMI approach, in *Nonlinear Analysis: Modelling and Control*, vol. 19, no. 2 (2014), pp. 256–271
56. C.C. Lin, C.C. Chang, J.F. Wang, Active control of irregular buildings considering soil–structure interaction effects, in *Soil Dynamics and Earthquake Engineering*, vol. 30, no. 3 (2010), pp. 98–109
57. T.H. Nguyen, N.M. Kwok, Q.P. Ha, J. Li, B. Samali, Adaptive sliding mode control for civil structures using magnetorheological dampers, in *International Symposium on Automation and Robotics in Construction* (2006)
58. K. Iwamoto, K. Yuji, K. Nonami, K. Tanida, I. Iwasaki, Output feedback sliding mode control for bending and torsional vibration control of 6-story flexible structure. JSME Int. J. Ser. C **45**(1), 150–158 (2002)
59. J. Ghaboussi, A. Joghataie, Active control of structures using neural networks. J. Eng. Mech. **121**(4), 555–567 (1995)
60. X. Jiang, H. Adeli, Pseudospectra, MUSIC, and dynamic wavelet neural network for damage detection of highrise buildings. Int. J. Numer. Methods Eng. **71**(5), 606–629 (2007)
61. K. Bani-Hani, J. Ghaboussi, Nonlinear structural control using neural networks. J. Eng. Mech. **24**(3), 319–327 (1998)
62. J.T. Kim, H.J. Jung, I.W. Lee, Optimal structural control using neural networks. J. Eng. Mech. **126**(2), 201–205 (2000)
63. S. Suresh, S. Narasimhan, S. Nagarajaiah, Direct adaptive neural controller for the active control of nonlinear base-isolated buildings, in *Structural Control and Health Monitoring*, vol. 19, no. 3 (2011), pp. 370–384
64. N.D. Lagaros, V. Plevris, M. Papadrakakis, Neurocomputing strategies for solving reliability-robust design optimization problems. Eng. Comput. 27(7), 819–840 (2010)
65. N.D. Lagaros, M. Fragiadakis, Fragility assessment of steel frames using neural networks. Earthq. Spectra **23**(4), 735–752 (2007)
66. N.D. Lagaros, M. Papadrakakis, Neural network based prediction schemes of the non-linear seismic response of 3D buildings. Adv. Eng. Softw. **44**(1), 92–115 (2012)
67. J. Wang, C. Zhang, H. Zhu, X. Huang, L. Zhang, RBF Nonsmooth control method for vibration of building structure with actuator failure. Complexity **2017,** Article ID 2513815, 7 p (2017)
68. L.A. Zadeh, Fuzzy sets. Inf. Control **8**(3), 338–353 (1965)
69. D.A. Shook, P.N. Roschke, P.Y. Lin, C.H. Loh, Semi-active control of a torsionally-responsive structure. Eng. Struct. **31**(1), 57–68 (2009)
70. D.A. Shook, P.N. Roschke, P.Y. Lin, C.H. Loh, GA-optimized fuzzy logic control of a large-scale building for seismic loads. Eng. Struct. **30**(2), 436–449 (2008)
71. D.G. Reigles, M.D. Symans, Supervisory fuzzy control of a base-isolated benchmark building utilizing a neuro-fuzzy model of controllable fluid viscous dampers, in *Structural Control and Health Monitoring*, vol. 13, no. 2–3 (2006), pp. 724–747
72. H. Adeli, X. Jiang, Dynamic fuzzy wavelet neural network model for structural system identification. J. Struct. Eng. **132**(1), 102–111 (2006)
73. J.H. Holland, *Adaptation in Natural and Artificial Systems* (MIT Press, 1975)
74. C. Camp, S. Pezeshk, G. Cao, Optimized design of two-dimensional structures using a genetic algorithm. J. Struct. Eng. **124**(5), 551–559 (1998)
75. H.N. Li, X.L. Li, Experiment and analysis of torsional seismic responses for asymmetric structures with semi-active control by MR dampers. Smart Mater. Struct. **18**(7) (2009)
76. X. Jiang, H. Adeli, Neuro-genetic algorithm for non-linear active control of structures. Int. J. Numer. Methods Eng. **75**(7), 770–786 (2008)

77. O. Yoshida, S.J. Dyke, Response control of full-scale irregular buildings using magnetorheological dampers. J. Struct. Eng. **131**(5), 734–742 (2005)
78. H.N. Li, Z.G. Chang, G.B. Song, D.S. Li, Studies on structural vibration control with MR dampers using GA, in *American Control Conference,* vol. 6, Boston, MA (2004), pp. 5478–5482
79. Y. Arfiadi, M.N.S. Hadi, Passive and active control of three-dimensional buildings. Earthq. Eng. Struct. Dyn. **29**(3), 377–396 (2000)
80. W.A. Crossley, A.M. Cook, D.W. Fanjoy, V.B. Venkayya, Using the two branch tournament genetic algorithm for multiobjective design. AIAA J. **37**(2), 261–267 (1999)
81. H.-N. Li, L.-S. Huo, Optimal design of liquid dampers for torsionally coupled vibration of structures. Intell. Control Autom. **5**, 4535–4538 (2004)

Chapter 2
Structure Models in Bidirection

This chapter provides an overview of modeling of building structures under bidirectional earthquakes. Structural mechanics involves the study of vibrations incorporated in structures. In order to control a structure effectively, it is important to have the knowledge about its dynamics. The control of structures is associated with the safeguard of building structures from unidirectional or bidirectional seismic forces. One of the structural design objects is to model dynamic loadings and to produce innovative approach to curb vibration. The vibration control generates the required dynamics in the building structures within a stable range. This control design is decided by the structure of mathematical model [1, 2]. In [3], a compact relationship between the controller and the structure model is established. All engineering structures are composed of intrinsic mass and elastic characteristics. The dynamic modeling has similar characteristics with the static analysis. However, the dynamic analysis is much complex than static analysis. For example, the mass modeling technique for the dynamic model requires an elastic model and a mass model minutely refined by discrete masses [4].

2.1 Bidirectional Excitation

Recent earthquakes exhibit that the bidirectional effect is the main damage source of the structural damage. The seismic analysis should consider the bidirectional excitation. Generally, the earthquake exhibits arbitrary direction which is represented as bidirectional ground movement, and it could reduce the participation of the traverse frames to the structure torsional and lateral stiffness. A noteworthy change in the elastic torsional behavior of the building is observed considering a nonlinear behavior in the transverse frames. The effect of the magnitude of the axial forces acting in the corner columns in case of bidirectional ground motion subjected to structures is different from that in case of unidirectional ground motion [5]. In [6], it was suggested

W. Yu and S. Paul, *Active Control of Bidirectional Structural Vibration*,
SpringerBriefs in Applied Sciences and Technology,
https://doi.org/10.1007/978-3-030-46650-3_2

that for a structure exposed to two simultaneous horizontal earthquake components, the transverse element behavior can be nonlinear and so the contribution to the real torsional stiffness is smaller. In [7], the analysis of one-storey models with and without transverse elements subjected to unidirectional and bidirectional earthquakes was presented. Their study concluded that the addition of the transverse elements in the model significantly hampers the response of the border elements when the structure is subjected to the bidirectional seismic waves. The analysis of real buildings suggests that it is asymmetric in nature to some degree with a formal symmetric plan. The asymmetric nature of building will induce lateral as well as torsional vibrations simultaneously and is termed as torsion coupling (TC) considering the case of pure translational excitations. Soil–structure interaction (SSI) effects are considered and can be significant in case of the building structures constructed on soft medium. The effects of SSI can critically modify the dynamic characteristics of a structure such as natural frequencies, damping ratios, and mode shapes [8]. The knowledge of behavior and impact of the excitation forces play a significant role in the formulation of the building structures dynamic model. The movement of the portion of the earth crust is termed as earthquake which is accompanied by the sudden release of stresses. Usually, the epicenters for earthquake exist <25 miles below the earth's surface and are followed by a series of vibrations. The bidirectional ground forces exerting on the building structure are shown in Fig. 2.1. These forces result in series of structure vibrations.

The forces acting on the x- and y-axes can be illustrated by the following dynamic equations:

$$f_x = -m\ddot{x}_g \quad f_y = -m\ddot{y}_g, \tag{2.1}$$

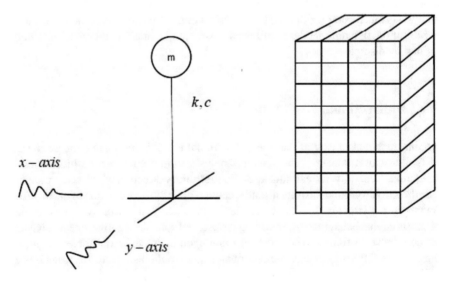

Fig. 2.1 Bidirectional ground forces exert on the building structures

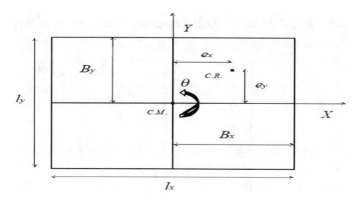

Fig. 2.2 The torsion coupled force

where m is the mass, $\ddot{x}_g$ and $\ddot{y}_g$ are the ground accelerations, caused by the seismic waves.

The main factors of the seismic movement for the building are the amplitude (displacement, velocity, and acceleration) and the frequency of the ground motion. The ground motion is complex, and the vibration frequency is time-varying. The ground motion and the building vibration affect each other, depending on the distance between the natural frequency of the building structure and seismic motion frequency. When the seismic wave frequency is close to the natural frequency of the building, the damages become bigger. Structure analysis shows that the shorter the building, the higher the natural frequency. One of the prime concerns is to control the structure vibrating with respect to low frequency, because the major part of the structure elastic energy is stored in low-frequency zone [9].

A controllable building structure can be regarded as a planar structure on a fixed base. The asymmetric characteristic of the building induces simultaneous lateral and torsional vibration, known as torsion coupling (TC) [10], which are subjected to bidirectional seismic inputs. The schematic plan view of structure involving TC is shown in Fig. 2.2. The impacts of seismic forces in x- and y-directions result in building oscillation as shown in Fig. 2.3. In includes x-, y-, and the torsional oscillations defined as θ.

The simplest structure is a one-storey under lateral translational motion at the roof level. It is a single degree of freedom system. The motion model is [11]

$$m\ddot{v} + c\dot{v} + kv = p(t), \tag{2.2}$$

where m is the mass, c is the damping, k is the stiffness, $\ddot{v}$ is the acceleration of the mass, $\dot{v}$ is the velocity of the mass relative to the base, v is the displacement, and $p(t)$ is the applied force, see Fig. 2.4.

Simple sway oscillations   **Combined sway and torsional oscillations**

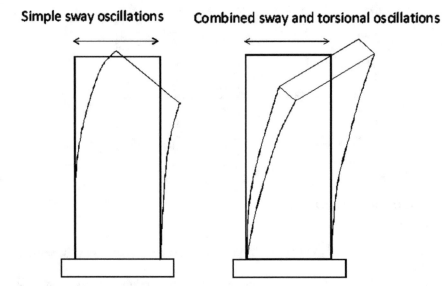

Fig. 2.3 The seismic forces result in building oscillation

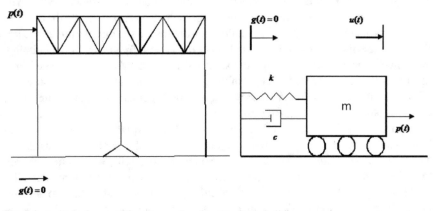

Fig. 2.4 A single degree of freedom system for one-storey building

Similarly, the equation of motion of a linear structure with n-degree of freedom (n-DOF) can be expressed as

$$M\ddot{X} + C\dot{X} + KX = P(t), \tag{2.3}$$

where M, C, and $K \in \mathbb{R}^{n \times n}$ are the mass, damping, and stiffness matrices, respectively, $\ddot{X}$, $\dot{X}$, and $X \in \mathbb{R}^{n \times 1}$ are the relative acceleration, velocity, and displacement vectors, respectively, and $P(t) \in \mathbb{R}^{n \times 1}$ is the external force vector.

The technique of modeling the stiffness parameter K can be on the basis of either a linear (elastic) or a nonlinear (inelastic) component [12]. The linear case means the relationship between the lateral force and the resulting deformation is linear [13]. When both ground translation and rotation are considered, the motion equation is [14]

$$M\ddot{X} + C\dot{X} + KX = F - MI_n\ddot{a}_g, \tag{2.4}$$

where $\ddot{a}_g$ represents the earthquake acceleration component, I_n is the system influence coefficient vector, $X = [x^T, y^T, \theta^T]^T$, $x = [x_1, \cdots x_n]^T$, $y = [y_1, \cdots y_n]^T$, $\theta = [\theta_1, \cdots \theta_n]^T$, $I_n = [I_1 I_2 0]^T$, $\ddot{a}_g = [\ddot{x}_g \ddot{y}_g \, 0]$. The mathematical analysis of the TC structure yields the following mass matrix, damping matrix, and stiffness matrix:

$$M = \begin{bmatrix} M_x & 0_{nxn} & 0_{nxn} \\ 0_{nxn} & M_y & 0_{nxn} \\ 0_{nxn} & 0_{nxn} & J_0 \end{bmatrix}, \quad K = \begin{bmatrix} K_{xx} & 0_{nxn} & -K_{x\theta} \\ 0_{nxn} & K_{yy} & K_{y\theta} \\ -K_{x\theta} & K_{y\theta} & K_{\theta\theta} \end{bmatrix},$$

where $J_0 = \text{diag}\left[m_1 r_1^2, \cdots m_n r_n^2\right]$, J_0 is the polar moment of inertia of the storey, r is the radius of gyration of the floor, n is the number of stories of the building, and C is the damping matrix which is proportional to mass and stiffness matrix by the Rayleigh method [15].

For a simple case, the mass of each floor is concentrated at the floor plate (N-storey shear model). Two seismic waves are in the x-direction and the y-direction. Here, the torsional components are zero, see Fig. 2.5. The left figure represents three-dimensional building structures, and the right figure exhibits the parameters of each floor. The motion equations show the relative displacements of the building structures with respect to the ground motions [16]

$$\begin{aligned} m_j\ddot{x}_j + p_{j-1} - p_j &= -m_j\ddot{x}_g(t) \\ m_j\ddot{y}_j + q_{j-1} - q_j &= -m_j\ddot{y}_g(t) \\ J_j\ddot{\theta}_j + r_{j-1} - r_j &= 0, \end{aligned} \tag{2.5}$$

where x_j and y_j are the jth floor displacement in x- and y-directions, respectively, and θ_j is the jth floor torsion angle relative to the ground. p_{j-1} and q_{j-1} are the jth floor column shear forces in x- and y-directions, p_j and q_j are the $j + 1$th floor column shear forces in x- and y-directions, respectively, r_{j-1} is the jth floor torque generated by the shear forces, r_j is the $j + 1$th floor torque, m_j is the mass of the jth floor, and J_j is the rotational inertia. In the above motion equation, the $\ddot{x}_g(t)$ and $\ddot{y}_g(t)$ are the ground accelerations that strikes the building structures due to an earthquake. The total forces exerted on the each floor of the buildings in x- and y-directions are multiplied by the mass of the building at each floor. The torsional component of the ground acceleration is neglected and so the right-hand side of the third equation is zero. The movement of the buildings is in x- and y-directions, that is, the acceleration components are $\ddot{x}_j$ and $\ddot{y}_j$, respectively. Due to the bidirectional

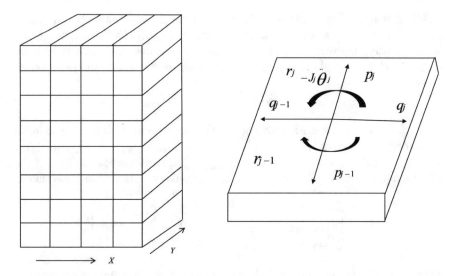

Fig. 2.5 Three-dimensional building structures with parameters of each floor

motion of the building, there will be coupling action on the building which give rise to the torsional motion in the building which is denoted by the component $\ddot{\theta}_j$.

If we only consider x-axis seismic wave, the torsion effect on the building is in x-component [17], see Fig. 2.6. The motion equations are

$$
\begin{aligned}
p_{j-1} &= K_j(x_j - x_{j-1}) + C_j(\dot{x}_j - \dot{x}_{j-1}) + B_j(\theta_j - \theta_{j-1}) + D_j(\dot{\theta}_j - \dot{\theta}_{j-1}) \\
r_{j-1} &= B_j(x_j - x_{j-1}) + D_j(\dot{x}_j - \dot{x}_{j-1}) + E_j(\theta_j - \theta_{j-1}) + F_j(\dot{\theta}_j - \dot{\theta}_{j-1}) \\
r_j &= B_{j+1}(x_{j+1} - x_j) + D_{j+1}(\dot{x}_{j+1} - \dot{x}_j) + E_{j+1}(\theta_{j+1} - \theta_j) + F_{j+1}(\dot{\theta}_{j+1} - \dot{\theta}_j),
\end{aligned}
\tag{2.6}
$$

where $K_j = \sum_{i=1}^{I} K_{j,i}$, $C_j = \sum_{i=1}^{I} C_{j,i}$, $B_j = \sum_{i=1}^{I} K_j, il_{j,i}$, $D_j = \sum_{i=1}^{I} C_j, il$, $E_j = \sum_{i=1}^{I} K_j, il_{i,j}^2$, $F_j = \sum_{i=1}^{I} C_{i,j}l_{i,j}^2$, K_j, i, and C_j, i are the stiffness and viscous damping coefficient, respectively, of the ith plane frame at the jth floor, m_j is the mass of the jth floor, J_j is the moment of inertia of the jth floor, $l_{j,i}$ is the distance of mass center of the jth floor to the ith plane frame, and I is total number of plane frames. $l_{j,i}$ is positive if the ith plane frame is located on the left of the mass center; otherwise, it is negative.

2.2 Structure Model of a Two-Floor Building

The normal method of structure design regards the seismic response arising from the ground motion that acts separately in the two orthogonal directions. Generally, the earthquake exhibits arbitrary direction which is represented as bidirectional ground

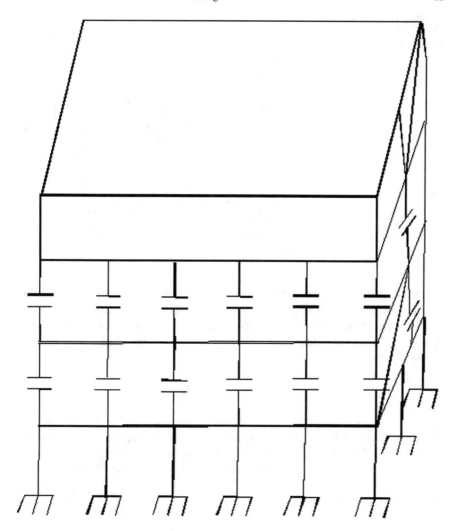

Fig. 2.6 The torsion effect on the building is in x-component

movement, and it could reduce the participation of the traverse frames to the structure torsional and lateral stiffness, see Fig. 2.1. The simplest structure is a one-storey building; for one direction, it can be modeled by [18],

$$m\ddot{x} + c\dot{x} + s\,x = f_e, \tag{2.7}$$

where m is the mass, c is the damping coefficient, s is the stiffness, f_e is an external force applied to the structure, and x, $\dot{x}$, and $\ddot{x}$ are the displacement, velocity, and acceleration, respectively.

Fig. 2.7 A two-floor
building

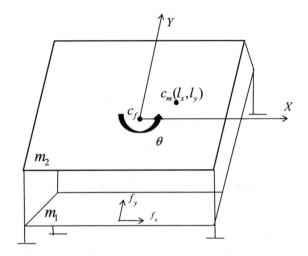

If the external force is in bidirectional, there are not only vibrations in X- and Y-axes, but also torsion coupling. The torsional oscillation comes from the asymmetric characteristic of the building, i.e., the physical center (c_f) is different with the mass center (c_m), see the two-floor building in Fig. 2.7. The motion of an n-floor structure can be expressed as [13, 19]

$$M\ddot{\mathbf{x}} + C\dot{\mathbf{x}} + \mathbf{f}_s = \mathbf{f}_e, \tag{2.8}$$

where $\mathbf{x} \in \Re^n$, $M \in \Re^{n \times n}$, $C \in \Re^{n \times n}$, $f_s = \begin{bmatrix} f_{s,1} \cdots f_{s,n} \end{bmatrix} \in \Re^n$ is the structure stiffness force vector, and $f_e \in \Re^n$ is the external force vector applied to the structure.

Also $M = \text{diag}\left(M_x, M_y, J_t\right) \in \Re^{(3n) \times (3n)}$, diag $(\cdot)$ is a diagonal matrix, $M_x = M_y = \text{diag}\left(m_1 \cdots m_n\right)$, m_i is the mass of the ith floor, $J_t = \text{diag}\left(m_1 r_1^2 \cdots m_n r_n^2\right)$ is the polar moment of inertia. $f_e = \begin{bmatrix} f_x, f_y \end{bmatrix}^T = \begin{bmatrix} -M_x & 0 \\ 0 & -M_y \\ 0 & 0 \end{bmatrix} \begin{bmatrix} a_x \\ a_y \end{bmatrix}$, where a_x are a_y are the accelerations of the external force in X- and Y-directions. The displacements of the structure, with respect to the bidirectional force $f_e = \begin{bmatrix} f_x, f_y \end{bmatrix}^T$, have 3 components $x = [x, y, \theta]^T$, and θ is the torsional angle.

The structure stiffness force f_s can be modeled as a linear model or a nonlinear model. In simple linear case

$$\mathbf{f}_s = S\mathbf{x}, \tag{2.9}$$

where $x = [x_1 \cdots x_n, y_1 \cdots y_n, \theta_1 \cdots \theta_n]^T \in \Re^{3n}$,

$$S = \begin{bmatrix} S_x & 0 & -S_{x\theta} \\ 0 & S_y & S_{y\theta} \\ -S_{x\theta} & S_{y\theta} & S_\theta \end{bmatrix}, \quad S_\rho = \begin{bmatrix} s_{\rho_1} + s_{\rho_2} & -s_{\rho_2} & \cdots & 0 & 0 \\ \vdots & \vdots & \ddots & \vdots & \vdots \\ 0 & 0 & \cdots & -s_{\rho_n} & s_{\rho_n} \end{bmatrix}, \quad \rho = (x, y, \theta)$$

$s_{\theta i} = s_{\theta_i} + s_{x_i} l_{y_i}^2 + s_{y_i} l_{x_i}^2, \quad i = 1 \cdots n$ represents floor, s is the stiffness

$$S_{x\theta} = \begin{bmatrix} s_{x_1} l_{y_1} + s_{x_2} l_{y_2} & -s_{x_2} l_{y_2} & \cdots & 0 & 0 \\ \vdots & \vdots & \ddots & \vdots & \vdots \\ 0 & 0 & \cdots & -s_{x_n} l_{y_n} & s_{x_n} l_{y_n} \end{bmatrix} \in \Re^{3n}$$

$$S_{y\theta} = \begin{bmatrix} s_{y_1} l_{x_1} + s_{y_2} l_{x_2} & -s_{y_2} l_{x_2} & \cdots & 0 & 0 \\ \vdots & \vdots & \ddots & \vdots & \vdots \\ 0 & 0 & \cdots & -s_{y_n} l_{x_n} & s_{y_n} l_{x_n} \end{bmatrix} \in \Re^{3n}.$$

The matrix S represents the overall stiffness matrix, whereas the matrix S_ρ represents the stiffness matrix in x-, y-, and θ-directions, respectively, by substituting ρ. Also, l_{x_i} and l_{y_i} represents the length of the structure in X- and Y-directions, respectively, $i = 1, 2$. The overall stiffness matrix S can be calculated by substituting S_x, S_y, and S_θ using S_ρ and also by substituting $S_{x\theta}$ and $S_{y\theta}$.

The damping matrix C is proportional to mass matrix M and stiffness matrix S

$$C = aM + bS,$$

using Rayleigh method, represented by the equation. Therefore, the damping matrix C has the same form as of the stiffness matrix S, $C = \begin{bmatrix} C_x & 0 & -C_{x\theta} \\ 0 & C_y & C_{y\theta} \\ -C_{x\theta} & C_{y\theta} & C_\theta \end{bmatrix}$.

2.3 Nonlinear Stiffness

When the structure is under the grip of very strong force which deforms the structure beyond its limit of linear elastic behavior, the structure stiffness force f_s cannot be modeled as a linear model. The behavior of the structure can be demonstrated using Bouc–Wen model. The advantages incorporated in the Bouc–Wen model are that it can demonstrate inelastic behavior of the structure where the strength/stiffness degradation can be easily incorporated. The relationship between the forces and displacements is [20]

$$f_{\rho,i} = \varepsilon_{\rho i} s_{\rho i} x_{\rho i} + (1 - \varepsilon_{\rho i}) s_{\rho i} \phi_{\rho i}, \tag{2.10}$$

where $\rho = (x, y)$, $i = 1 \cdots n$, $\varepsilon_{\rho i}$ are positive numbers.

The first part of (2.10) is the elastic stiffness, and the second part is the inelastic stiffness. The nonlinear function $\phi_{\rho i}$ is

$$\phi_{\rho i} = \frac{1}{\eta_{\rho i}}[A\dot{x}_{\rho i} - \beta_{\rho i}\,|\dot{x}_{\rho i}|\,|a_{\rho i}|^{\eta_i-1}\,a_{\rho i}v_{\rho i} + \gamma_{\rho i}\,|\dot{x}_{\rho i}|\,|a_{\rho i}|^{\eta_i-1}\,v_{\rho i}a_{\rho i}\,\mathrm{sign}(\dot{x}_{\rho i}a_{\rho i})],$$

(2.11)

where A, $\beta_{\rho i}$, $\gamma_{\rho i}$, $\alpha_{\rho i}$, n, and η are positive numbers.

$\eta_{\rho i} = 1 + \delta_{\rho i}E_{\rho i}$ controls the stiffness degradation, and $v_{\rho i} = 1 + \delta_{\rho i}E_{\rho i}$ controls strength degradation. Normalized dissipated hysteretic energy is

$$E_{\rho i} = (1 - \alpha_{\rho i})\int_0^t \frac{\dot{x}_{\rho i}a_{\rho i}}{\Delta_{\rho i}\Delta_{\rho i}}dt \quad \Delta_{\rho i} = (\beta_{\rho i} + \gamma_{\rho i})^{-\frac{1}{\eta\rho i}}.$$

(2.12)

The property of passivity states that the system storage energy is always lesser than the energy supplied. In [21], it was demonstrated that the Bouc–Wen model is considered to be passive with respect to its energy storage. The nonlinear differential equation (2.11) is continuous, and also it is dependent on time. The property of local Lipschitz is also maintained. It can be validated that (2.11) has a unique solution on a time interval $[0, t_0]$. In the stability analysis involved in the other chapters, this property will be utilized.

2.4 Conclusion

In this chapter, the importance of modeling of building structures under bidirectional earthquakes is highlighted. It is important to have the knowledge of the model dynamics for effective implementation of the control law. The modeling equations of the AMD and TA are laid down.

References

1. A. Forrai, S. Hashimoto, H. Funato, K. Kamiyama, Structural control technology: system identification and control of flexible structures. Comput. Control Eng. J. **12**(6), 257–262 (2001)
2. J. Zhang, P.N. Roschke, Active control of a tall structure excited by wind. J. Wind Eng. Indus. Aerodyn. **83**(1–3), 209–223 (1999)
3. G.W. Housner, L.A. Bergman, T.K. Caughey, A.G. Chassiakos, R.O. Claus, S.F. Masri, R.E. Skeleton, T.T. Soong, B.F. Spencer Jr., J.T.P. Yao, Structural control: past, present and future. J. Eng. Mech. **123**(9), 897–971 (1997)
4. B. Donaldson, *Introduction to Structural Dynamics* (Cambridge University Press, UK, 2006)
5. E. Cruz, S. Cominetti, Three-Dimensional buildings subjected to bidirectional earthquakes. Validity of analysis considering unidirectional earthquakes, in *12th World Conference on Earthquake Engineering* (2000)
6. W.K. Tso, T.J. Zhu, Design of torsionally unbalanced structural systems based on code provisions I: ductility demands. Earthq. Eng. Struct. Dyn. **21**(7) (1992)

7. J.C. Correnza, G.L. Hutchinson, A.M. Chandler, Effect of transverse load-resisting elements on inelastic earthquake response of eccentric-plan buildings. Earthq. Eng. Struct. Dyn. **23**(1), 75–89 (1994)
8. W.H. Wu, J.F. Wang, C.C. Lin, Systematic assessment of irregular building–soil interaction using efficient modal analysis. Earthq. Eng. Struct. Dyn. **30**(4), 573–594 (2001)
9. K.M. Choi, S.W. Cho, D.O. Kim, I.W. Lee, Active control for seismic response reduction using modal-fuzzy approach. Int. J. Solids Struct. **42**(16–17), 4779–4794 (2005)
10. C.C. Lin, C.-C. Chang, J.F. Wang, Active control of irregular buildings considering soil–structure interaction effects. Soil Dyn. Earthq. Eng. **30**(3), 98–109 (2010)
11. J.C. Anderson, F. Naeim, *Basic Structural Dynamics* (Wiley, Los Angeles, CA, 2012)
12. A.C. Nerves, R. Krishnan, Active control strategies for tall civil structures, in *International Conference on Industrial Electronics, Control, and Instrumentation,* vol. 2, Orlando, FL (1995), pp. 962–967
13. A.K. Chopra, *Dynamics of Structures*. Prentice-Hall International Series (2011)
14. F.Y. Cheng, *Matrix Analysis of Structural Dynamics: Applications and Earthquake Engineering* (CRC Press, New York, 2000)
15. G.C. Hart, K. Wong, *Structural Dynamics for Structural Engineers* (Wiley, 1999)
16. J. Zhang, K. Zeng, J. Jiang, An optimal design of bi-directional TMD for three dimensional structure. Comput. Struct. Eng. 935–941 (2009)
17. B. Wu, J.P. Ou, T.T. Soong, Optimal placement of energy dissipation devices for three-dimensional structures. Eng. Struct. **19**(2), 113–125 (1997)
18. A. Alavinasab, H. Moharrami, A. Khajepour, Active control of structures using energy-based LQR method, in *Computer-Aided Civil and Infrastructure Engineering,* vol. 21, no. 8 (2006), pp. 605–611
19. A.B.M. Saiful Islam, R.R. Hussain, M. Jameel et al., Non-linear time domain analysis of base isolated multi-storey building under site specific bi-directional seismic loading. Autom. Construct. **22**, 554–566 (2012)
20. C.S. Lee, H.P. Hong, Statistics of inelastic responses of hysteretic systems under bidirectional seismic excitations. Eng. Struct. **32**(8), 2074–2086 (2010)
21. F. Ikhouanea, V. Mañosaa, J. Rodellara, Dynamic properties of the hysteretic Bouc-Wen model. Syst. Control Lett. **56**(3), 197–205 (2007)

Chapter 3
Bidirectional PD/PID Control of Structures

3.1 Introduction

The role of the structural control is to minimize the vibrations of the buildings under
the effect of bidirectional earthquake via an effective external control force. In an
active control system, it is essential to design an effective control strategy, which
is simple, robust, and fault-tolerant. Several attempts have been made to implement
advanced controllers for the active vibration control of structures as discussed in
Chap. 1. Chapter 1 clearly justifies that the control device plays a superior role in
preventing structure from damages. A good control law gives good performance
of the anti-vibration. The pole-placement H_∞ control with target damping ratio is
proposed by [1]. In [2], the genetic algorithm is applied to determine the feedback
control. Many optimal control algorithms are applied for the active vibration control of structures, for example, filtered linear quadratic control [3], linear quadratic
regulator (LQR) [4], and linear quadratic Gaussian (LQG) control [5]. The active
mass damper (AMD) is widely implemented, which utilizes the mass without spring
and dashpot [6]. Due to the existence of translation–torsion coupled vibrations with
respect to the bidirectional seismic waves, in this chapter, a torsional actuator (TA) is
utilized. It is a disk-motor device, which is incorporated in the structure to minimize
the torsional response of the building. For real application, an effective controller
should be simple, robust, and fault-tolerant. The PD/PID control has been widespread
applied in industrial processes. It may be the best control, because it shows its effectivity without the knowledge of the model and also due to its simplicity as well
as it is incorporated with distinct physical meanings. In [7], MR damper is utilized to control a three-dimensional structure from bidirectional seismic excitations.
The controller used the mechanism of PD control. It calculates the essential forces
required to control the structural vibrations. In [8], PD and PID controllers are used
in the numerical simulations to control structure under unidirectional earthquake.
Nigdeli and Boduroglu [9] used active tendons to control torsionally irregular and

multi-storey structures under the effect of near-fault ground motion excitation. In their work, PID-type controllers were used to generate the control signals. In [10], various feedback control strategies in relation to active control of earthquake utilizing PID-type controllers were presented. A numerical algorithm was taken into consideration for finding out the parameters of PID. An effective technique involving synergistic combination of PID controller and LQR methodology for seismic control of structures is proposed in [11]. The problems of existed bidirectional PD/PID control are as follows: (1) they do not consider the lateral–torsional control mechanism, that is, only AMD was used to mitigate the lateral–torsional vibration but a combination of AMD and TA is not implemented; (2) they do not analyze the stability of closed-loop system.

In this chapter, standard PD and PID controls are utilized as active vibration control of the structure in order to solve the above two problems. Initially, the analysis is based on the lateral–torsional vibration, linear and nonlinear structure stiffness, and the hysteresis of the structure model under the bidirectional wave. Then, the sufficient conditions for asymptotic stability of the PD/PID control are validated by utilizing Lyapunov stability analysis. These conditions are quite convenient for the designer to choose the controller gains straightaway. An active vibration control system with two floors equipped with an AMD and a TA is set up in order to carry out the experimental analysis. The experimental results using the PD and PID controllers validate their effectiveness and stability.

3.2 Active Control of Structural Vibration

In order to minimize the vibrations caused by the bidirectional external forces (f_x and f_y), an AMD and a TA are installed on the structure as shown in Fig. 3.1. AMD is placed near the mass center of the building. The TA is placed on the physical center of the building.

The control force is $u = \left[u_x, u_y, u_\theta\right]^T$. Considering the building model (2.8) in the previous chapter and the control, the closed-loop system is

$$M\ddot{x} + C\dot{x} + \mathbf{f}_s - \mathbf{f}_e = \Gamma(\mathbf{u} - \mathbf{d_u}). \tag{3.1}$$

The closed-loop system represented by (3.1) is the control equation that is utilized for control and stability analysis. In the forthcoming section, this equation is subdivided into three components mainly X-, Y-, and θ-components, and then analysis is carried out. In (3.1), $u \in \mathfrak{R}^{3n}$ is the control signals which is fed to the dampers, where dampers are AMD and TA in combination, $d_\mathbf{u}$ is the damping and friction force vector of the dampers, and Γ is the location matrix of the dampers which is defined as

$$\Gamma_{i,j} = \begin{cases} 1 \text{ if } i = j = f_l \\ 0 \quad \text{otherwise} \end{cases}, \tag{3.2}$$

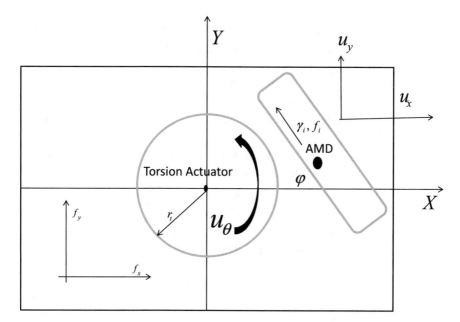

Fig. 3.1 Bidirectional active control of structures

where $\forall i, j \in \{1, \ldots, n\}$, $f_l \subseteq \{1, \ldots, n\}$, f_l are the floors on which the dampers are installed. For a two-floor building, $\Gamma = \begin{bmatrix} \Gamma_{1,1} & \Gamma_{1,2} \\ \Gamma_{2,1} & \Gamma_{2,2} \end{bmatrix}$. If the damper is placed on the second floor, $f_l = \{2\}$, $\Gamma = \begin{bmatrix} 0 & 0 \\ 0 & 1 \end{bmatrix}$. If the damper is placed on both first and second floors, $f_l = \{1, 2\}$, $\Gamma = \begin{bmatrix} 1 & 0 \\ 0 & 1 \end{bmatrix}$.

If we illustrate the closed-loop system mentioned by (3.1) along all three directions, that is, X-, Y-, and θ-directions, then

$$M_x \ddot{x} + C_x \dot{x} + f_{sx} - f_x = \Gamma (u_x - d_{ux})$$
$$M_y \ddot{y} + C_y \dot{y} + f_{sx} - f_y = \Gamma (u_y - d_{uy}) \qquad (3.3)$$
$$J_t \ddot{\theta} + C_\theta \dot{\theta} + f_{s\theta} = \Gamma (u_\theta - d_{u\theta}).$$

The AMD force in ith floor is defined as f_i

$$f_i = m_{di}(\ddot{d}_i + \ddot{\gamma}_i), \qquad (3.4)$$

where m_{di} is the mass of the AMD, $\ddot{d}_i$ is the acceleration of the AMD, $\ddot{\gamma}_i$ is the acceleration of the structure along the AMD, $\ddot{\gamma}_i = \sqrt{a_{i,x}^2 + a_{i,y}^2}$. f_i should be separated into X- and Y-directions as

$$u_{i,x} = f_i \cos \varphi = m_{di}(\ddot{d}_i \cos \varphi + a_{i,x})$$
$$u_{i,y} = f_i \sin \varphi = m_{di}(\ddot{d}_i \sin \varphi + a_{i,y})$$
$$\ddot{\gamma}_i = \frac{a_{i,x}}{\cos \varphi} = \frac{a_{i,y}}{\sin \varphi}$$
$$\ddot{x}_{i,x} = a_{i,x} + \ddot{d}_i \cos \varphi$$
$$\ddot{x}_{i,y} = a_{i,y} + \ddot{d}_i \sin \varphi,$$

where φ is the angle of the AMD along X-axis, and $\ddot{x}_{i,x}$ and $\ddot{x}_{i,y}$ are the relative accelerations of the AMD along X- and Y- directions. So

$$f_i = m_{di}\left(\ddot{d}_i + \frac{a_{i,x}}{\cos \varphi}\right) = m_{di}\left(\ddot{d}_i + \frac{a_{i,y}}{\sin \varphi}\right).$$

We define the control force of the AMD along X- and Y-directions as $u_d = [u_x, u_y]^T$

$$u_{di} = m_{di}\left[\ddot{x}_{i,x}, \ddot{x}_{i,y}\right]^T.$$

Consider the friction of the AMD,

$$fr_{i,x} = c\dot{x}_{i,x} + \varepsilon m_{di}g \tanh\left[\beta\dot{x}_{i,x}\right]$$
$$fr_{i,y} = c\dot{x}_{i,y} + \varepsilon m_{di}g \tanh\left[\beta\dot{x}_{i,y}\right],$$

where c, β, and ε are the damping coefficients of the Column friction [12]. The final control of the AMD is

$$
\begin{aligned}
u_x &= m_{di}\ddot{x}_{i,x} - fr_{i,x} \\
u_y &= m_{di}\ddot{x}_{i,y} - fr_{i,y}.
\end{aligned}
\tag{3.5}
$$

Using (3.3) and (3.5),

$$
\begin{aligned}
M_x\ddot{x} + C_x\dot{x} + f_{sx} - f_x &= \Gamma(m_{di}\ddot{x}_{i,x} - fr_{i,x} - d_{ux}) \\
M_y\ddot{y} + C_y\dot{y} + f_{sx} - f_y &= \Gamma(m_{di}\ddot{x}_{i,y} - fr_{i,y} - d_{uy}).
\end{aligned}
\tag{3.6}
$$

The third element of the control $u = [u_x, u_y, u_\theta]^T$ is the torsion force u_θ. The TA is a rotating disk equipped with DC motor and is placed at the center of physical center, see Fig. 3.1. The control object is to decrease the torsional response of the structures due to the bidirectional movements and the mass center and the physical center being different. The inertia moment of TA is

$$J_t = m_t r_t^2,
\tag{3.7}$$

where m_t is the mass of the disk and r_t is the radius of the disk. The torque τ generated by the disk is

$$u_\theta = J_t(\ddot{\theta}_t + \ddot{\theta}),
\tag{3.8}$$

where $\ddot{\theta}$ is the angular acceleration of the building, and $\ddot{\theta}_t$ is the angular acceleration of the TA. Obviously, to decrease the torsional response, the directions of $\ddot{\theta}_t$ and $\ddot{\theta}$ should be different. Consider the friction of the TA

$$fr_t = c\dot{\theta}_t + F_c \tanh(\beta\dot{\theta}_t), \tag{3.9}$$

where c is the torsional viscous friction coefficient, F_c is the coulomb friction torque, and tanh is the hyperbolic tangent depending on β and motor speed. The final torsion control is

$$u_\theta = J_t(\ddot{\theta}_t + \ddot{\theta}) - fr_t. \tag{3.10}$$

Using (3.3) and (3.10),

$$J_t\ddot{\theta} + C_\theta\dot{\theta} + f_{s\theta} = \Gamma(J_t(\ddot{\theta}_t + \ddot{\theta}) - fr_t - d_{u\theta}). \tag{3.11}$$

The main role of the AMD is to reduce the response of acceleration of building in X- and Y-directions, whereas the main role of the TA is to minimize the torsional effect on the building. For the closed-loop system (3.1)

$$\mathbf{d_u} = \begin{bmatrix} c\dot{x}_{i,x} + \varepsilon m_{di}g \tanh\left[\beta\dot{x}_{i,x}\right] = d_{ux} \\ c\dot{x}_{i,y} + \varepsilon m_{di}g \tanh\left[\beta\dot{x}_{i,y}\right] = d_{uy} \\ c\dot{\theta}_t + F_c \tanh(\beta\dot{\theta}_t) = d_{u\theta} \end{bmatrix}. \tag{3.12}$$

The movements of the AMD and TA are sliding in nature. The sliding mechanism of the actuators absorbs the energy from the friction. The kinetic energy is converted into heat energy in this phenomenon. So the coefficients in the friction models are assumed to be Coulomb.

3.3 PD/PID Controller of Building Structures

As PD control is very simple and robust to uncertainties, it is the most popular controller for mechanical systems. It is the simplest controller for the structural vibration control system. PD controller is the best choice. PD control has the following form when the AMD control is coupled with the TA control:

$$\mathbf{u} = -K_p\mathbf{e} - K_d\dot{\mathbf{e}}, \tag{3.13}$$

where $e = x - x^d$, $x = [x, y, \theta]^T$, x^d is desired reference vector, for the vibration control, and $x^d = 0$. K_p and K_d are positive-definite constant matrices that correspond to the proportional and derivative gains. The PD control (3.13) for structures becomes

$$\mathbf{u} = -K_p\mathbf{x} - K_d\dot{\mathbf{x}}. \tag{3.14}$$

The design of the controller is based on the suitable gain selection K_p and K_d in (3.13), such that the closed-loop system is stable and good performances are achieved. For the bidirectional structure control, the gains of one-floor PD are as follows: $K_P = \text{diag}\left(K_{px}, K_{py}, K_{p\theta}\right) \in \Re^{6\times6}$, $K_d = \text{diag}\left(K_{dx}, K_{dy}, K_{d\theta}\right)$. The closed-loop system (3.1) with the PD control shown in (3.14) is

$$M\ddot{\mathbf{x}} + C\dot{\mathbf{x}} + S\mathbf{x} + \mathbf{f}_e + \Gamma\mathbf{d}_\mathbf{u} = \Gamma(-K_p\mathbf{x} - K_d\dot{\mathbf{x}}). \qquad (3.15)$$

Here the terms $(S\mathbf{x} + \mathbf{f}_e + \Gamma\mathbf{d}_\mathbf{u})$ can be regarded as uncertainties. In the following section, we assume that it satisfies the Lipschitz condition.

It is well known that the regulation error becomes smaller while increasing the derivative gain. The cost of large derivative gain results in slow transient performance. Only when derivative gain tends to infinity, the regulation error converges to zero [13]. However, it would seem better to use a smaller derivative gain if the system contains high-frequency noise signals.

From the control viewpoint, the regulation error can be removed by introducing an integral component to the PD control. PID controllers use feedback strategy and have three actions. P action is introduced for increasing the speed of response. D action is introduced for damping purposes. I action is introduced for obtaining a desired steady-state response [14]. The PID control is

$$\mathbf{u} = -K_p(\mathbf{x} - \mathbf{x}^d) - K_i \int_0^t (\mathbf{x} - \mathbf{x}^d)d\tau - K_d(\dot{\mathbf{x}} - \dot{\mathbf{x}}^d), \qquad (3.16)$$

where K_p, K_i, and K_d are positive-definite, and K_i is the integration gain. For the structure control, $x^d = \dot{x}^d = 0$ (3.16) becomes

$$\mathbf{u} = -K_p\mathbf{x} - K_i \int_0^t \mathbf{x}d\tau - K_d\dot{\mathbf{x}}. \qquad (3.17)$$

In order to analyze PID controller, (3.17) is expressed by

$$\mathbf{u} = -K_p\mathbf{x} - K_d\dot{\mathbf{x}} - \boldsymbol{\xi}$$
$$\boldsymbol{\xi} = K_i \int_0^t \mathbf{x}d\tau, \quad \boldsymbol{\xi}(0) = \mathbf{0}.$$

The closed-loop system (3.1) with the PID control (3.16) becomes

$$M\ddot{\mathbf{x}} + C\dot{\mathbf{x}} + S\mathbf{x} + \mathbf{f}_e + \Gamma\mathbf{d}_\mathbf{u} = -K_{px}\mathbf{x} - K_{dx}\dot{\mathbf{x}} - \boldsymbol{\xi} \qquad (3.18)$$
$$\dot{\boldsymbol{\xi}} = K_i\mathbf{x}.$$

In matrix form, the closed-loop system is

$$\frac{d}{dt}\begin{bmatrix} \boldsymbol{\xi} \\ \mathbf{x} \\ \dot{\mathbf{x}} \end{bmatrix} = \begin{bmatrix} K_{ix}\mathbf{x} \\ \dot{\mathbf{x}} \\ -M^{-1}\left(C\dot{\mathbf{x}} + S\mathbf{x} + \mathbf{f}_e + \Gamma\mathbf{d}_\mathbf{u} + K_{px}\mathbf{x} + K_{dx}\dot{\mathbf{x}} + \boldsymbol{\xi}\right) \end{bmatrix}. \qquad (3.19)$$

Unlike the H_2 control [15] and optimal control [4], PD control does not need the model of the structure. The model discussed in the above section will be used for stability analysis in this chapter. The theory analysis of bidirectional PD control is still not appeared in publications [8].

3.4 Stability Analysis

As the combined forces generated by AMD and TA are fed to the structure, this force may stabilize or destabilize the structure. If the control algorithm generates unstable signal, the AMD and TA will generate forces that can make the structure unstable. This matter becomes more complicated for nonlinear devices, as a bounded input signal may also make nonlinear devices to generate unstable output.

In general cases, the structures associated with open-loop systems are asymptotically stable. This is true for the case when there is no external force, $f_e = 0$. The criteria are valid in case of inelastic stiffness because of its BIBO stability and passivity properties. In the event of seismic excitation, the ideal active control force required for canceling out the vibration completely is $\Gamma u = f_e$. But practically it is not possible because f_e is not always measurable and is much bigger than any control device force. Therefore, the main intention of the active control is to maintain the vibration as minimum as possible by mitigating the relative movement between the structural floors. Normally, the structural parameters are partly known, and the structure model might have embedded nonlinearity such as the hysteresis phenomenon.

It is quite favorable to represent the closed-loop system (3.15) with PD control as

$$M\ddot{x} + C\dot{x} + f = -\Gamma \left(K_p x + K_d \dot{x} \right), \tag{3.20}$$

where $f = f_s + f_e + d_u$.

The following theorem gives the stability analysis of the PD control (3.14). To simplify the proof, we assume that $\Gamma_{n \times n} = I_{n \times n}$, which means that each floor has an AMD and a TA.

Theorem 3.1 *Consider the structural system as (3.1) controlled by the PD controller as (3.14), the closed-loop system as (3.15) is stable, provided that the control gains are positive. The regulation errors converge to the following residual sets:*

$$
\begin{aligned}
D_x &= \left\{ \mathbf{x}, \dot{} \mid \|\dot{\mathbf{x}}\|_{Q_x}^2 + \left\| \dot{} \right\|_{Q_x}^2 \leq \bar{\mu}_{\mathbf{fx}} + \tfrac{2}{\mathbf{f}} \bar{\mu}_{\mathbf{fx}} \right\} \\
D_y &= \left\{ \mathbf{y}, \dot{} \mid \|\dot{\mathbf{y}}\|_{Q_y}^2 + \left\| \dot{} \right\|_{Q_y}^2 \leq \bar{\mu}_{\mathbf{fy}} + \tfrac{2}{\mathbf{f}} \bar{\mu}_{\mathbf{fy}} \right\} \\
D_\theta &= \left\{ \dot{}, \dot{\mathbf{x}}, \dot{\mathbf{y}} \mid \left\| \dot{} \right\|_{Q_\theta}^2 + \|\dot{\mathbf{x}}\|_{Q_\theta}^2 + \|\dot{\mathbf{y}}\|_{Q_\theta}^2 \leq \bar{\mu}_{\mathbf{f}_\theta} + \bar{\mathbf{f}}^{-2} \bar{\mu}_{\mathbf{fx}} + \bar{\mathbf{f}}^{-2} \bar{\mu}_{\mathbf{fy}} \right\},
\end{aligned}
\tag{3.21}
$$

where $\bar{\mu}_{fx} \geq f_x^T \Lambda_f^{-1} f_x$, $\bar{\mu}_{fy} \geq f_y^T \Lambda_f^{-1} f_y$, $\bar{\mu}_f \geq f^T \Lambda_f^{-1} f$, $C_x + \alpha_f C_{x\theta} > \Lambda_{fx} > 0$, $C_y + \eta_f C_{y\theta} > \Lambda_{fy} > 0$, $C_\theta + \alpha_f^{-1} C_{x\theta} + \eta_f^{-1} C_{y\theta} > \Lambda_f > 0$.

Proof The closed-loop system (3.20) can also be represented as

$$
\begin{bmatrix} M_x \ddot{x} \\ M_y \ddot{y} \\ J_t \ddot{\theta} \end{bmatrix} + \begin{bmatrix} C_x \dot{x} - C_{x\theta} \dot{\theta} \\ C_y \dot{y} + C_{y\theta} \dot{\theta} \\ C_\theta \dot{\theta} - C_{x\theta} \dot{x} + C_{y\theta} \dot{y} \end{bmatrix} + \begin{bmatrix} \mathbf{f_x} \\ \mathbf{f_y} \\ \mathbf{f} \end{bmatrix} = -\Gamma \left\{ \begin{bmatrix} K_{px} x \\ K_{py} y \\ K_{p\theta} \theta \end{bmatrix} + \begin{bmatrix} K_{dx} \dot{x} \\ K_{dy} \dot{y} \\ K_{d\theta} \dot{\theta} \end{bmatrix} \right\}.
\tag{3.22}
$$

Therefore, we have three sets to represent X-, Y-, and θ-directions

$$
\begin{aligned}
M_x \ddot{x} + \left(C_x \dot{x} - C_{x\theta} \dot{\theta} \right) + \mathbf{f}_x &= -\Gamma(K_{px} x + K_{dx} \dot{x}) \\
M_y \ddot{y} + \left(C_y \dot{y} + C_{y\theta} \dot{\theta} \right) + \mathbf{f}_y &= -\Gamma(K_{py} y + K_{dy} \dot{y}) \\
J_t \ddot{\theta} + \left(C_\theta \dot{\theta} - C_{x\theta} \dot{x} + C_{y\theta} \dot{y} \right) + \mathbf{f} &= -\Gamma(K_{p\theta} \theta + K_{d\theta} \dot{\theta}).
\end{aligned}
\tag{3.23}
$$

We will analyze one by one. Since the AMD and TA are placed in the second floor, $\Gamma = 1$. If all the individual structural equations in (3.23) with PD controller are stable, the entire structural is stable using PD controller. For that purpose, we select Lyapunov candidate as

$$
V_x = \frac{1}{2} \dot{x}^T M_x \dot{x} + \frac{1}{2} x^T K_{px} x.
\tag{3.24}
$$

The first term of (3.24) signifies the kinetic energy, and the second term denotes elastic potential energy. As M_x and K_{px} are positive-definite matrices, $V_x \geq 0$. The derivative of (3.24) is

$$
\begin{aligned}
\dot{V}_x &= \dot{x}^T M_x \ddot{x} + \dot{x}^T K_{px} x \\
&= \dot{x}^T \left(-C_x \dot{x} + C_{x\theta} \dot{\theta} - \mathbf{f}_x - K_{px} x - K_{dx} \dot{x} \right) + \dot{x}^T K_{px} x \\
&= -\dot{x}^T \left(C_x + K_{dx} \right) \dot{x} + \dot{x}^T C_{x\theta} \dot{\theta} - \dot{x}^T \mathbf{f}_x.
\end{aligned}
\tag{3.25}
$$

Using the matrix inequality,

$$
X^T Y + Y^T X \leq X^T \Lambda X + Y^T \Lambda^{-1} Y.
\tag{3.26}
$$

It is valid for any $X, Y \in \Re^{n \times m}$ and any $0 < \Lambda = \Lambda^T \in \Re^{n \times n}$; we can write the scalar variable $\dot{x}^T f_x$ as

$$
\dot{x}^T \mathbf{f}_x = \frac{1}{2} \dot{x}^T \mathbf{f}_x + \frac{1}{2} \mathbf{f}_x^T \dot{x} \leq \dot{x}^T \Lambda_{fx} \dot{x} + \mathbf{f}^T \Lambda_{fx}^{-1} \mathbf{f},
\tag{3.27}
$$

where Λ_{fx} is any positive-definite matrix. Now $\dot{x}$ and $\dot{\theta}$ are related to each other as the vibration along x-direction will create a torsional movement θ and so we suppose

$$\dot{\theta} = -\alpha_{\mathbf{f}}\dot{x}, \tag{3.28}$$

where $\alpha_{\mathbf{f}}$ is a positive-definite matrix. As the X component of ground acceleration will give a torsion in the structure in anti-clockwise sense, hence, we assumed the relation to be negative. Using (3.28) in (3.25),

$$\dot{V}_x = -\dot{x}^T (C_x + K_{dx})\dot{x} - \alpha_{\mathbf{f}}^T \dot{x} C_{x\theta}\dot{x} - \dot{x}^T \mathbf{f}_x$$
$$\dot{V}_x = -\dot{x}^T (C_x + {}_{\mathbf{f}}C_{x\theta} + K_{dx})\dot{x} - \dot{x}^T \mathbf{f}_x.$$

We select $\Lambda_{\mathbf{f}x}$ as

$$C_x + {}_{\mathbf{f}}C_{x\theta} > \Lambda_{\mathbf{f}x} > 0. \tag{3.29}$$

So

$$\dot{V}_x \leq -\dot{x}^T (C_{xx} + {}_{\mathbf{f}}C_{x\theta} + K_{dx} - \Lambda_{\mathbf{f}x})\dot{x} + \mathbf{f}_x^T \Lambda_{\mathbf{f}x}^{-1}\mathbf{f}_x. \tag{3.30}$$

If we choose the gain $K_{dx} > 0$, and also since $\alpha_{\mathbf{f}}$ is positive-definite matrix, $C_{xx} > 0$, $C_{x\theta} > 0$, we have

$$\dot{V}_x \leq -\dot{x}^T Q_x \dot{x} + \bar{\mu}_{\mathbf{f}x} \leq -\lambda_m (Q_x) \|\dot{x}\|^2 + \mathbf{f}_x^T \Lambda_{\mathbf{f}x}^{-1}\mathbf{f}_x, \tag{3.31}$$

where $Q_x = C_x + \alpha_{\mathbf{f}}C_{x\theta} + K_{dx} - \Lambda_{\mathbf{f}x} > 0$. $\dot{V}_x$ is therefore an ISS-Lyapunov function. Using Theorem 3.1 from [16], the boundedness of $f_x^T \Lambda_{\mathbf{f}x}^{-1} f_x \leq \bar{\mu}_{\mathbf{f}x}$ implies that the regulation error $\|\dot{x}\|$ is bounded,

$$\|\dot{x}\|_{Q_x}^2 > \bar{\mu}_{\mathbf{f}\dot{x}}, \quad \forall t \in [0, T], \tag{3.32}$$

then we can conclude that $\dot{V}_x < 0$ when $\|\dot{x}\|_{Q_x}^2 > \bar{\mu}_{\mathbf{f}\dot{x}}$. From (3.28), we have

$$\dot{x} = -{}_{\mathbf{f}}^{-1}\dot{\theta}, \quad |\dot{x}| = -{}_{\mathbf{f}}^{-1}|\dot{\theta}|, \quad |\dot{x}||\dot{x}| = -{}_{\mathbf{f}}^{-1}|\dot{\theta}||\dot{x}|, \quad \|\dot{x}\|^2 = {}_{\mathbf{f}}^{-2}\|\dot{\theta}\|^2. \tag{3.33}$$

Implementing (3.33) in (3.32), we have

$$\|\dot{\theta}\|_{Q_x}^2 > {}_{\mathbf{f}}^2 \bar{\mu}_{\mathbf{f}x}, \forall t \in [0, \Upsilon]. \tag{3.34}$$

Above condition also satisfies $\dot{V}_x < 0$ when $\|\dot{\theta}\|_{Q_x}^2 > \alpha_{\mathbf{f}}^2 \bar{\mu}_{\mathbf{f}x}$. Adding (3.32) and (3.34), we have

$$\|\dot{x}\|_{Q_x}^2 + \|\dot{\theta}\|_{Q_x}^2 > \bar{\mu}_{\mathbf{f}x} + {}_{\mathbf{f}}^2 \bar{\mu}_{\mathbf{f}x}, \forall t \in [0, T + \Upsilon]. \tag{3.35}$$

Now we show that the total time during which $\|\dot{x}\|_{Q_x}^2 + \|\dot{\theta}\|_{Q_x}^2 > \bar{\mu}_{\mathbf{f}x} + \alpha_{\mathbf{f}}^2 \bar{\mu}_{\mathbf{f}x}$ is finite. Let T_k denote the time interval during which $\|\dot{x}\|_{Q_x}^2 + \|\dot{\theta}\|_{Q_x}^2 > \bar{\mu}_{\mathbf{f}x} + \alpha_{\mathbf{f}}^2 \bar{\mu}_{\mathbf{f}x}$. $\|\dot{x}\|_{Q_x}^2 + \|\dot{\theta}\|_{Q_x}^2 > \bar{\mu}_{\mathbf{f}x} + \alpha_{\mathbf{f}}^2 \bar{\mu}_{\mathbf{f}x}$ will stay inside the circle in case $\|\dot{x}\|_{Q_x}^2 + \|\dot{\theta}\|_{Q_x}^2 >$

$\bar{\mu}_{\mathrm{f}x} + \alpha_{\mathrm{f}}^2 \bar{\mu}_{\mathrm{f}x}$ stay outside the circle of radius $\bar{\mu}_{\mathrm{f}x} + \alpha_{\mathrm{f}}^2 \bar{\mu}_{\mathrm{f}x}$ for finite times and then re-enter the circle. Also, $\sum_{k=1}^{\infty} T_k < \infty$, since the total time $\|\dot{x}\|_{Q_x}^2 + \|\dot{\theta}\|_{Q_x}^2 > \bar{\mu}_{\mathrm{f}x} + \alpha_{\mathrm{f}}^2 \bar{\mu}_{\mathrm{f}x}$ is finite and

$$\lim_{k \to \infty} T_k = 0. \tag{3.36}$$

So $\|\dot{x}\|_{Q_x}^2 + \|\dot{\theta}\|_{Q_x}^2$ is bounded via an invariant set argument. Also using (3.31) it can be shown that $\|\dot{x}\|$ and $\left\|\dot{}\right\|$ are also bounded. Let $\|\dot{x}\|_{Q_x}^2 + \|\dot{\theta}\|_{Q_x}^2$ denote the largest tracking error during the T_k interval. Then using (3.36) and bounded $\|\dot{x}\|_{Q_x}^2 + \|\dot{\theta}\|_{Q_x}^2$ imply that

$$\lim_{k \to \infty} \left[\|\dot{x}\|_{Q_x}^2 + \|\dot{\theta}\|_{Q_x}^2 - (\bar{\mu}_{\mathrm{f}x} + {}_{\mathrm{f}}^2 \bar{\mu}_{\mathrm{f}x}) \right] = 0.$$

So $\|\dot{x}\|_{Q_x}^2 + \|\dot{\theta}\|_{Q_x}^2$ will converge to $\bar{\mu}_{\mathrm{f}x} + \alpha_{\mathrm{f}}^2 \bar{\mu}_{\mathrm{f}x}$. Therefore, the derivative of regulation error x and θ converges to the residual set

$$\dot{D}_x = \left\{ \dot{x}, \dot{\theta} \mid \|\dot{x}\|_{Q_x}^2 + \|\dot{\theta}\|_{Q_x}^2 \leq \bar{\mu}_{\mathrm{f}x} + {}_{\mathrm{f}}^2 \bar{\mu}_{\mathrm{f}x} \right\}. \tag{3.37}$$

Also for $\|\dot{x}\|_{Q_x}^2 > \bar{\mu}_{\mathrm{f}x}$, the total time is finite and hence $V_x = \frac{1}{2}\dot{x}^T M_x \dot{x} + \frac{1}{2}x^T K_{px} x$ is bounded; hence, the regulation error $\dot{x}$ is bounded. Also for $\|\dot{\theta}\|_{Q_x}^2 > \alpha_{\mathrm{f}}^2 \bar{\mu}_{\mathrm{f}x}$, the total time is finite and hence assuming $V_\theta = \frac{1}{2}\dot{\theta}^T J_t \dot{\theta} + \frac{1}{2}\theta^T K_{p\theta}\theta$ it can be shown to be bounded and so regulation error $\dot{\theta}$ is also bounded, $\dot{\theta} = -\alpha_{\mathrm{f}}\dot{x}$ and $J_t = M_x r^2$. Again using the Lyapunov candidate $V_y = \frac{1}{2}\dot{y}^T M_y \dot{y} + \frac{1}{2}y^T K_{py} y$, and using the similar sort of stability analysis, we can infer that the derivative of regulation error y and θ converges to the residual set

$$\dot{D}_y = \left\{ \dot{y}, \dot{\theta} \mid \|\dot{y}\|_{Q_y}^2 + \|\dot{\theta}\|_{Q_y}^2 \leq \bar{\mu}_{\mathrm{f}y} + {}_{\mathrm{f}}^2 \bar{\mu}_{\mathrm{f}y} \right\}, \tag{3.38}$$

where

$$\dot{\theta} = {}_{\mathrm{f}}\dot{y} \tag{3.39}$$

is positive due to clockwise sense

$$Q_y = C_y + {}_{\mathrm{f}}C_{y\theta} + K_{dy} - \Lambda_{\mathrm{f}y} > 0.$$

For $\|\dot{y}\|_{Q_y}^2 > \bar{\mu}_{\mathrm{f}y}$, the total time is finite and hence $V_y = \frac{1}{2}\dot{y}^T M_y \dot{y} + \frac{1}{2}y^T K_{py} y$ is bounded; hence, the regulation error $\dot{y}$ is bounded. Also for $\|\dot{\theta}\|_{Q_y}^2 > \alpha_{\mathrm{f}}^2 \bar{\mu}_{\mathrm{f}y}$, the total time is finite, and hence assuming $V_\theta = \frac{1}{2}\dot{\theta}^T J_0 \dot{\theta} + \frac{1}{2}\theta^T K_{p\theta}\theta$, it can be shown to be bounded and so regulation error $\dot{\theta}$ is also bounded, $\dot{\theta} = \eta_{\mathrm{f}}\dot{y}$ and $J_t = M_y r^2$.

Using the Lyapunov candidate $V_\theta = \frac{1}{2}\dot{\theta}^T J_0 \dot{\theta} + \frac{1}{2}\theta^T K_{p\theta}\theta$, and using the similar sort of stability analysis, we can infer that the derivative of regulation error x, y, and θ converges to the residual set

$$D_\theta = \left\{ \dot{\theta}, \dot{x}, \dot{y} \mid \|\dot{\theta}\|^2_{Q_\theta} + \|\dot{x}\|^2_{Q_\theta} + \|\dot{y}\|^2_{Q_\theta} \leq \bar{\mu}_{f_\theta} + \bar{f}^{-2}\bar{\mu}_{fx} + \bar{f}^{-2}\bar{\mu}_{fy} \right\}, \quad (3.40)$$

where $\dot{x} = -\alpha_f^{-1}\dot{\theta}$, $\dot{y} = \eta_f^{-1}\dot{\theta}$, $Q_\theta = C_\theta + \alpha_f^{-1}C_{x\theta} + \eta_f^{-1}C_{y\theta} + K_{d\theta} - \Lambda_f > 0$. For $\|\dot{\theta}\|^2_{Q_\theta} > \bar{\mu}_{f_\theta}$, the total time is finite, and hence assuming $V_\theta = \frac{1}{2}\dot{\theta}^T J_0 \dot{\theta} + \frac{1}{2}\theta^T K_{p\theta}\theta$ it can be shown to be bounded and so regulation error $\dot{\theta}$ is also bounded. For $\|\dot{x}\|^2_{Q_\theta} > \alpha_f^{-2}\bar{\mu}_{fx}$, the total time is finite and hence $V_x = \frac{1}{2}\dot{x}^T M_x \dot{x} + \frac{1}{2}x^T K_{px}x$ is bounded; hence, the regulation error $\dot{x}$ is bounded. For $\|\dot{y}\|^2_{Q_\theta} > \eta_f^{-2}\bar{\mu}_{fy}$, the total time is finite and hence $V_y = \frac{1}{2}\dot{y}^T M_y \dot{y} + \frac{1}{2}y^T K_{py}y$ is bounded; hence, the regulation error $\dot{y}$ is bounded.

From the above section, it is clear that in order to decrease the regulation errors caused by these uncertainties, the derivative gain K_d has to be increased but will result in slow response. Now we analyze the stability of the bidirectional PID control (3.17). The equilibrium of (3.19) is $[\boldsymbol{\xi}_x, x, \dot{x}] = [\boldsymbol{\xi}_x^*, \boldsymbol{0}, \boldsymbol{0}]$. Since at equilibrium point $x = 0$ and $\dot{x} = 0$, the equilibrium is $[\mathbf{f}(0), \boldsymbol{0}, \boldsymbol{0}]$. In order to move the equilibrium to origin, we define

$$\boldsymbol{\xi}_x = \boldsymbol{\xi}_x - \mathbf{f}(0). \quad (3.41)$$

The final closed-loop equation becomes

$$M_x \ddot{x} + \left(C_x + {}_fC_{x\theta}\right)\dot{x} + \mathbf{f}_x = -K_{px}x - K_{dx}\dot{x} - \boldsymbol{\xi}_x + \mathbf{f}(0)$$
$$\boldsymbol{\xi}_x = K_{ix}x. \quad (3.42)$$

In a similar manner,

$$M_y \ddot{y} + \left(C_y + {}_fC_{y\theta}\right)\dot{y} + \mathbf{f}_y = -K_{py}y - K_{dy}\dot{y} - \boldsymbol{\xi}_y + \mathbf{f}(0)$$
$$\boldsymbol{\xi}_y = K_{iy}y \quad (3.43)$$
$$J_0 \ddot{\theta} + \left(C_\theta + {}_f^{-1}C_{x\theta} + {}_f^{-1}C_{y\theta}\right)\dot{\theta} + \mathbf{f} = -K_{p\theta}\theta - K_{d\theta}\dot{\theta} - \boldsymbol{\xi}_\theta + \mathbf{f}(0)$$
$$\boldsymbol{\xi}_\theta = K_{i\theta}\theta.$$

In order to analyze the stability of (3.42) and (3.43), we need the following properties.

P1. The positive-definite matrix $M = M_x = M_y$ satisfies the following Condition:

$$0 < \lambda_m(M) \leq \|M\| \leq \lambda_M(M) \leq \bar{m}$$
$$0 < \lambda_{j_t}(J_t) \leq \|J_t\| \leq \lambda_{J_t}(J_t) \leq \bar{j}_t,$$

where $\lambda_m(M)$ and $\lambda_M(M)$ are the minimum and maximum eigenvalues of the matrix M, respectively, and $\bar{m} > 0$ is the upper bound, $\lambda_{j_t}(J_t)$ and $\lambda_{J_t}(J_t)$ are the minimum and maximum eigenvalues of the matrix J_t, respectively, and $\bar{j}_t > 0$ is the upper bound.

P2. The term f is Lipschitz over $\tilde{x}$ and $\tilde{y}$

$$\|\mathbf{f}(\tilde{x}) - \mathbf{f}(\tilde{y})\| \leq k_{\mathbf{f}} \|\tilde{x} - \tilde{y}\|. \tag{3.44}$$

Most of the uncertainties are first-order continuous functions. Since f_s, f_e, and $d_\mathbf{u}$ are first-order continuous functions and satisfy Lipschitz condition, P2 can be established.

We calculate the lower bound of $\int f\, dx$ as

$$\int_0^t \mathbf{f}\, dx = \int_0^t \mathbf{f}_s dx + \int_0^t \mathbf{f}_e dx + \int_0^t \mathbf{d_u}\, dx. \tag{3.45}$$

Here, we define the lower bound of $\int_0^t f_s dx$ as $-\bar{f}_s$, and $\int_0^t d_\mathbf{u}\, dx$ as $-\bar{d}_u$.

Compared with f_s and $d_\mathbf{u}$, f_e is much bigger in the case of earthquake. We define the lower bound of $\int_0^t f_e dx$ as $-\bar{f}_e$. Finally, the lower bound $k_{\mathbf{fx}}$ is

$$k_{\mathbf{fx}} = -\bar{f}_s - \bar{f}_e - \bar{d}_u. \tag{3.46}$$

The following theorem gives the stability analysis of PID controller.

Theorem 3.2 *Considering the structural system as (3.1) controlled by the PID controller as (3.17), the closed-loop systems (3.42) and (3.43) are asymptotically stable at the equilibriums* $\left[\boldsymbol{\xi}_x - \mathbf{f}(0), x, \dot{x}\right]^T = 0$, $\left[\boldsymbol{\xi}_y - \mathbf{f}(0), y, \dot{y}\right]^T = 0$ *and* $\left[\boldsymbol{\xi}_\theta - \mathbf{f}(0), \theta, \dot{\theta}\right]^T = 0$, *provided that the PID gains satisfy*

$$\lambda_m(K_{dx}) \geq \frac{1}{4}\sqrt{\frac{1}{3}\lambda_m(M_x)\lambda_m(K_{px})}\left[1 + \frac{k_{c_x} + \alpha_f k_{c_{x\theta}}}{\lambda_M(M_x)}\right]$$
$$\qquad - \lambda_m(C_x) - \lambda_m(\alpha_f C_{x\theta})$$
$$\lambda_M(K_{ix}) \leq \frac{1}{6}\sqrt{\frac{1}{3}\lambda_m(M_x)\lambda_m(K_{px})}\frac{\lambda_m(K_{px})}{\lambda_M(M_x)}$$
$$\lambda_m(K_{px}) \geq \frac{3}{2}[k_f + k_{c_x} + \alpha_f k_{c_{x\theta}}]$$

$$\lambda_m\left(K_{dy}\right) \geq \frac{1}{4}\sqrt{\frac{1}{3}\lambda_m\left(M_y\right)\lambda_m\left(K_{py}\right)}\left[1 + \frac{k_{c_y} + {}_\mathrm{f}k_{c_{y\theta}}}{\lambda_M\left(M_y\right)}\right]$$
$$- \lambda_m(C_y) - \lambda_m({}_\mathrm{f}C_{y\theta})$$

$$\lambda_M(K_{iy}) \leq \frac{1}{6}\sqrt{\frac{1}{3}\lambda_m\left(M_y\right)\lambda_m\left(K_{py}\right)}\frac{\lambda_m(K_{py})}{\lambda_M\left(M_y\right)}$$

$$\lambda_m(K_{py}) \geq \frac{3}{2}[k_f + k_{c_y} + {}_\mathrm{f}k_{c_{y\theta}}]$$

$$\lambda_{j_t}\left(K_{d\theta}\right) \geq \frac{1}{4}\sqrt{\frac{1}{3}\lambda_{j_t}\left(J_t\right)\lambda_{j_t}\left(K_{p\theta}\right)}\left[1 + \frac{k_{c_\theta} + \alpha_f^{-1}k_{c_{x\theta}} + {}_\mathrm{f}^{-1}k_{c_{y\theta}}}{\lambda_{J_t}\left(J_t\right)}\right]$$
$$- \lambda_{j_t}(C_\theta) - \lambda_{j_0}(\alpha_f^{-1}C_{x\theta}) - \lambda_{j_0}({}_\mathrm{f}^{-1}C_{y\theta})$$

$$\lambda_{J_t}(K_{i\theta}) \leq \frac{1}{6}\sqrt{\frac{1}{3}\lambda_{j_t}\left(J_t\right)\lambda_{j_t}\left(K_{p\theta}\right)}\frac{\lambda_{j_t}(K_{p\theta})}{\lambda_{J_t}\left(J_t\right)}$$

$$\lambda_{j_t}(K_{p\theta}) \geq \frac{3}{2}[k_f + k_{c_\theta} + \alpha_f^{-1}k_{c_{x\theta}} + {}_\mathrm{f}^{-1}k_{c_{y\theta}}].$$

Proof Here, the Lyapunov function is defined as

$$V_x = \frac{1}{2}\dot{x}^T M_x \dot{x} + \frac{1}{2}x^T K_{px}x + \frac{\sigma}{4}\xi_x^T K_{ix}^{-1}\xi_x + x^T \xi_x \\ + \frac{\sigma}{2}x^T M_x \dot{x} + \frac{\sigma}{4}x^T K_{dx}x + \int_0^t \mathbf{f}dx - k_{\mathbf{f}x}, \tag{3.47}$$

where $V(0) = 0$. In order to show that $V_x \geq 0$, it is separated into three parts, such that $V_x = V_{x1} + V_{x2} + V_{x3}$

$$V_{x1} = \frac{1}{6}x^T K_{px}x + \frac{\sigma}{4}x^T K_{dx}x + \int_0^t \mathbf{f}dx - k_{\mathbf{f}x} \geq 0, \; K_{px} > 0, \; K_{dx} > 0 \tag{3.48}$$

$$V_{x2} = \frac{1}{6}x^T K_{px}x + \frac{\sigma}{4}\xi_x^T K_{ix}^{-1}\xi_x + x^T \xi_x \\ \geq \frac{1}{2}\frac{1}{3}\lambda_m(K_{px})\|x\|^2 + \frac{\sigma\lambda_m(K_{ix}^{-1})}{4}\|\xi_x\|^2 - \|x\|\,\|\xi_x\|. \tag{3.49}$$

When $\sigma \geq \frac{3}{(\lambda_m(K_{ix}^{-1})\lambda_m(K_{px}))}$,

$$V_{x2} \geq \frac{1}{2}\left(\sqrt{\frac{\lambda_m(K_{px})}{3}}\|x\| - \sqrt{\frac{3}{4(\lambda_m(K_{px}))}}\|\xi_x\|\right)^2 \geq 0 \tag{3.50}$$

and

$$V_{3x} = \frac{1}{6}x^T K_{px} x + \frac{1}{2}\dot{x}^T M\dot{x} + \frac{\sigma}{2}x^T M\dot{x}. \tag{3.51}$$

Because

$$X^T A X \geq \|X\| \|AX\| \geq \|X\| \|A\| \|X\| \geq \lambda_M(A) \|X\|^2 \tag{3.52}$$

when

$$\sigma \leq \frac{1}{2}\frac{\sqrt{\frac{1}{3}\lambda_m(M_x)\lambda_m(K_{px})}}{\lambda_M(M_x)}$$

$$V_{x3} \geq \frac{1}{2}\left(\frac{1}{3}\lambda_m(K_{px})\|x\|^2 + \lambda_m(M_x)\|\dot{x}\|^2 + \sigma\lambda_M(M_x)\|x\|\|\dot{x}\|\right)$$

$$= \frac{1}{2}\left(\sqrt{\frac{\lambda_m(K_{px})}{3}}\|x\| + \sqrt{\lambda_m(M_x)}\|\dot{x}\|\right)^2 \geq 0. \tag{3.53}$$

Now, we have

$$\frac{1}{2}\frac{\sqrt{\frac{1}{3}\lambda_m(M_x)\lambda_m(K_{px})}}{\lambda_M(M_x)} \geq \sigma \geq \frac{3}{(\lambda_m(K_{ix}^{-1})\lambda_m(K_{px}))}. \tag{3.54}$$

The derivative of (3.47) is

$$\dot{V}_x = \dot{x}^T M_x\ddot{x} + \dot{x}^T K_{px}x + \frac{\sigma}{2}\boldsymbol{\xi}_x^T K_{ix}^{-1}\boldsymbol{\xi}_x + \dot{x}^T\boldsymbol{\xi}_x$$

$$+ x^T\boldsymbol{\xi}_x + \frac{\sigma}{2}\dot{x}^T M_x\dot{x} + \frac{\sigma}{2}x^T M_x\ddot{x} + \sigma\dot{x}^T K_{dx}x + \dot{x}^T\mathbf{f}. \tag{3.55}$$

Using (3.26), we can write

$$-\frac{\sigma}{2}x^T C_{xx}\dot{x} \leq \frac{\sigma}{2}k_{c_x}\left(x^T x + \dot{x}^T\dot{x}\right) \tag{3.56}$$

$$-\frac{\sigma\alpha_f}{2}x^T C_{x\theta}\dot{x} \leq \frac{\sigma\alpha_f}{2}k_{c_{x\theta}}\left(x^T x + \dot{x}^T\dot{x}\right),$$

where $\|C_x\| \leq k_{c_x}$ and $\|C_{x\theta}\| \leq k_{c_{x\theta}}$. So $\xi_x = K_{ix}$, $\xi^T K_{ix}^{-1}\xi$ becomes $x^T\xi$, and $x^T\xi$ becomes $x^T K_{ix}$. Using (3.56), we have

$$\dot{V}_x = -\dot{x}^T\left[C_x + \alpha_f C_{x\theta} + K_{dx} - \frac{\sigma}{2}M_x - \frac{\sigma}{2}k_{c_x} - \frac{\sigma\alpha_f}{2}k_{c_{x\theta}}\right]\dot{x}$$

$$- x^T\left[\frac{\sigma}{2}K_{px} - K_{ix} - \frac{\sigma}{2}k_{c_x} - \frac{\sigma\alpha_f}{2}k_{c_{x\theta}}\right]x - \frac{\sigma}{2}x^T[\mathbf{f}_x - \mathbf{f}(0)] + \dot{x}^T f(0). \tag{3.57}$$

Now using the Lipschitz condition (3.44)

$$\frac{\sigma}{2}x^T [\mathbf{f}(0) - \mathbf{f_x}] \leq \frac{\sigma}{2}k_\mathbf{f} \|x\|^2 \tag{3.58}$$

$$-\frac{\sigma}{2}x^T [\mathbf{f_x} - \mathbf{f}(0)] \leq x^T \frac{\sigma}{2}k_\mathbf{f}x.$$

From (3.26),

$$\dot{x}^T f(0) \geq -f^T(0)\Lambda^{-1T}f(0). \tag{3.59}$$

Using (3.58)

$$\dot{V}_x = -\dot{x}^T \left[C_x + \alpha_f C_{x\theta} + K_{dx} - \frac{\sigma}{2}M_x - \frac{\sigma}{2}k_{c_x} - \frac{\sigma\alpha_f}{2}k_{c_{x\theta}} \right]\dot{x}$$
$$- x^T \left[\frac{\sigma}{2}K_{px} - K_{ix} - \frac{\sigma}{2}k_{c_{xx}} - \frac{\sigma\alpha_f}{2}k_{c_{x\theta}} - \frac{\sigma}{2}k_f \right]x + \dot{x}^T f(0), \tag{3.60}$$

Equation (3.60) becomes

$$\dot{V}_x \leq -\dot{x}^T \left[\lambda_m(C_x) + \lambda_m(\alpha_f C_{x\theta}) + \lambda_m(K_{dx}) - \frac{\sigma}{2}\lambda_M(M_x) - \frac{\sigma}{2}k_{c_x} - \frac{\sigma\alpha_f}{2}k_{c_{x\theta}} \right]\dot{x}$$
$$- x^T \left[\frac{\sigma}{2}\lambda_m(K_{px}) - \lambda_M(K_{ix}) - \frac{\sigma}{2}k_{c_{xx}} - \frac{\sigma\alpha_f}{2}k_{c_{x\theta}} - \frac{\sigma}{2}k_f \right]x. \tag{3.61}$$

So $\dot{V}_x \leq 0$, $\|x\|$ minimizes if two conditions are met: (1) $\lambda_m(C_x) + \lambda_m(\alpha_f C_{x\theta}) + \lambda_m(K_{dx}) \geq \frac{\sigma}{2}[\lambda_M(M_x) + k_{c_x} + \alpha_f k_{c_{x\theta}}]$; (2) $\lambda_m(K_{px}) \geq \frac{2}{\sigma}\lambda_M(K_{ix}) + k_{c_x} + \frac{\sigma\alpha_f}{2}k_{c_{x\theta}} + k_f$. Now using (3.54) and $\lambda_m\left(K_{ix}^{-1}\right) = \frac{1}{\lambda_M(K_{ix})}$, we have

$$\lambda_m(K_{dx}) \geq \frac{1}{4}\sqrt{\frac{1}{3}\lambda_m(M_x)\lambda_m\left(K_{px}\right)}\left[1 + \frac{k_{c_x} + \alpha_f k_{c_{x\theta}}}{\lambda_M(M_x)}\right] - \lambda_m(C_x) - \lambda_m(\alpha_f C_{x\theta}) \tag{3.62}$$

again $\frac{2}{\sigma}\lambda_M(K_{ix}) = \frac{2}{3}\lambda_m(K_{px})$. Hence,

$$\lambda_M(K_{ix}) \leq \frac{1}{6}\sqrt{\frac{1}{3}\lambda_m(M_x)\lambda_m\left(K_{px}\right)}\frac{\lambda_m(K_{px})}{\lambda_M(M_x)}. \tag{3.63}$$

Also

$$\lambda_m(K_{px}) \geq \frac{3}{2}[k_f + k_{c_x} + \alpha_f k_{c_{x\theta}}]. \tag{3.64}$$

By the Lyapunov function

$$V_y = \frac{1}{2}\dot{y}^T M_y \dot{y} + \frac{1}{2}y^T K_{py}y + \frac{\sigma}{4}\xi_y^T K_{iy}^{-1}\xi_y + y^T \xi_y$$
$$+ \frac{\sigma}{2}y^T M_y \dot{y} + \frac{\sigma}{4}y^T K_{dy}y + \int_0^t \mathbf{f}dy - k_{\mathbf{f}y}. \tag{3.65}$$

Similarly, we can prove $V_y \geq 0$ and thus using stability analysis criteria we can prove $\dot{V}_y \leq 0$ if

$$\lambda_m\left(K_{dy}\right) \geq \frac{1}{4}\sqrt{\frac{1}{3}\lambda_m\left(M_y\right)\lambda_m\left(K_{py}\right)}\left[1 + \frac{k_{c_y} + \mathfrak{r}k_{c_{y\theta}}}{\lambda_M\left(M_y\right)}\right] - \lambda_m(C_y) - \lambda_m(\mathfrak{r}C_{y\theta})$$

$$\lambda_M(K_{iy}) \leq \frac{1}{6}\sqrt{\frac{1}{3}\lambda_m\left(M_y\right)\lambda_m\left(K_{py}\right)}\frac{\lambda_m(K_{py})}{\lambda_M\left(M_y\right)}$$

$$\lambda_m(K_{py}) \geq \frac{3}{2}[k_f + k_{c_y} + \mathfrak{r}k_{c_{y\theta}}]. \tag{3.66}$$

From Lyapunov function

$$V_\theta = \frac{1}{2}\dot{\theta}^T J_0\dot{\theta} + \frac{1}{2}\theta^T K_{p\theta}\theta + \frac{\sigma}{4}\xi\theta K_{i\theta}^{-1}\xi_\theta + \theta^T\xi_\theta + \frac{\sigma}{2}\theta^T J_0\dot{\theta}$$
$$+ \frac{\sigma}{4}\theta^T K_{dy}\theta + \int_0^t \mathfrak{f}d^{\hat{}} - k_{\mathfrak{f}\theta}. \tag{3.67}$$

Similarly, we can prove $V^{\cdot} \geq 0$. Thus, we can prove $\dot{V}^{\cdot} \leq 0$, $\|^{\hat{}}\|$ decreases if

$$\lambda_{j_t}\left(K_{d^{\hat{}}}\right) \geq \frac{1}{4}\sqrt{\frac{1}{3}\lambda_{j_t}\left(J_t\right)\lambda_{j_t}\left(K_{p\theta}\right)}\left[1 + \frac{k_{c_\theta} + \alpha_f^{-1}k_{c_{x\theta}} + \mathfrak{r}^{-1}k_{c_{y\theta}}}{\lambda_{J_t}\left(J_t\right)}\right.$$
$$\left. - \lambda_{j_t}(C_\theta) - \lambda_{j_t}(\alpha_f^{-1}C_{x\theta}) - \lambda_{j_t}(\mathfrak{r}^{-1}C_{y\theta})\right] \tag{3.68}$$

$$\lambda_{J_t}(K_{i\theta}) \leq \frac{1}{6}\sqrt{\frac{1}{3}\lambda_{j_t}\left(J_t\right)\lambda_{jt}\left(K_{p\theta}\right)}\frac{\lambda_{j_t}(K_{p\theta})}{\lambda_{J_t}\left(J_t\right)}$$

$$\lambda_{j_0}(K_{p\theta}) \geq \frac{3}{2}[k_f + k_{c_{\theta\theta}} + \alpha_f^{-1}k_{c_{x\theta}} + \mathfrak{r}^{-1}k_{c_{y\theta}}].$$

The above theorems suggest that the closed-loop system is asymptotically stable. But we cannot decide on the global stability of the closed-loop system. This is due to the fact that the hysteresis property is associated with the stiffness of the structure. The hysteresis output depends on the deformation factor all time. This deformation behaves according to the application or removal of forces. So the deformation is not the same before and after the application of forces, and hence the equilibrium position is also not static. Therefore, the equilibrium positions before and after the earthquake are not the same. The stable point gets shifted after an earthquake event.

Let us consider a ball of radius ς in the three-dimensional space. This three-dimensional space is represented by X-, Y-, and θ-components. The ball center is at origin of the state-space system where $\dot{V}_x \leq 0$, $\dot{V}_y \leq 0$, $\dot{V}_\theta \leq 0$. The origin of the closed-loop systems represented by (3.51), (3.52), and (3.53) are stable equilibrium. Now, we will prove for the asymptotic stability of the origin. For that, we use La Salle's theorem by defining the term Π_x, Π_y, and Π_θ as follows:

$$\Pi_x = \left\{ \bar{z}_x(t) = \left[x^T, \dot{x}^T, \xi_x^T \right]^T \in \Re^{3n} : \dot{V}_x = 0 \right\}, \xi_x \in \Re^n, x = 0 \in \Re^n, \dot{x} = 0 \in \Re^n$$

$$\Pi_y = \left\{ \bar{z}_y(t) = \left[y^T, \dot{y}^T, \xi_y^T \right]^T \in \Re^{3n} : \dot{V}_y = 0 \right\}, \xi_y \in \Re^n, y = 0 \in \Re^n, \dot{y} = 0 \in \Re^n$$

$$\Pi_\theta = \left\{ \bar{z}_\theta(t) = \left[\theta^T, \dot{\theta}^T, \xi_\theta^T \right]^T \in \Re^{3n} : \dot{V}_\theta = 0 \right\}, \xi_\theta \in \Re^n, \theta = 0 \in \Re^n, \dot{\theta} = 0 \in \Re^n.$$

$$(3.69)$$

Using (3.66) and substituting $x = 0$ and $\dot{x} = 0$, we have $\dot{V}_x = 0$. Similar analysis with $y = 0$ and $\dot{y} = 0$ and also $\theta = 0$ and $\dot{\theta} = 0$ will yield $\dot{V}_y = 0$ and $\dot{V}_\theta = 0$. Similarly, these conditions hold good for $\dot{x} = 0$, $\dot{y} = 0$ and $\dot{\theta} = 0$ for all $t \geq 0$. Therefore, $\bar{z}_x(t)$, $\bar{z}_y(t)$, and $\bar{z}_\theta(t)$ belongs to Π_x, Π_y, and Π_θ, respectively. Also, imparting these conditions to (3.51), (3.52), and (3.53), we have $\dot{\xi}_x = 0$, $\dot{\xi}_y = 0$, and $\dot{\xi}_\theta = 0$. Also, $\xi_x = 0$, $\xi_y = 0$, and $\xi_\theta = 0$ for all $t \geq 0$. So $\bar{z}_x(t)$ is the only initial condition in Π_x, $\bar{z}_y(t)$ is the only initial condition in Π_y, and $\bar{z}_\theta(t)$ is the only initial condition in Π_θ. Therefore, origin is asymptotically stable according to La Salle's theorem.

3.5 Experimental Results

In order to analyze and validate the bidirectional PD/PID controllers, a two-floor structure is designed and constructed as mentioned in the chapter experimental setup. This structure is then mounted on the shake table to carry out the experimental analysis which is illustrated in Fig. 3.2. The bidirectional shake table uses two Quanser one degree of freedom (I-40) [17] , which move in X- and Y-directions. The AMD [18] and TA are placed on the second floor of the structure. The total moving mass of the AMD and TA are taken to be the 5% of the total mass of the structure. The eccentricities of the physical center are 16 cm from X-direction and 27 cm from Y-direction. The eccentricities of the mass center are 15 cm from X-direction and 11 cm from Y-direction. The TA is placed on the physical center, whereas the AMD is placed on the mass center. The position of the AMD and the TA can be seen in Fig. 3.3.

The entire programming is carried out using the MATLAB and Simulink version R2011a. The Simulink program is used to generate the control actions for AMD and TA as well as for the movement of shake table. For creating a synchronization between MATLAB and Quanser devices, Quarc accelerate design version 2.3.603 is installed. The control actions between the computer and the dampers are synchronized using RT-DAC/USB data acquisition board [19]. The link between Simulink and RT-DAC/USB is achieved using RT CON toolbox which is provided with the hardware RT-DAC/USB. All the control actions were employed at a sampling frequency of 1 kHz. The biaxial accelerometers (XL403A) [20] are mounted on each floor.

Fig. 3.2 Bidirectional shake table

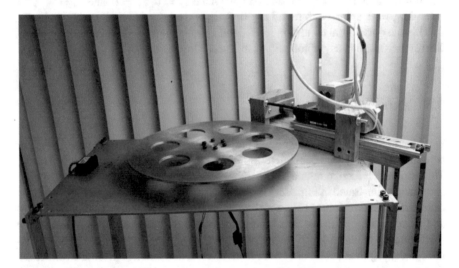

Fig. 3.3 Placement of AMD and TA

The acceleration of ground floor is subtracted from first floor acceleration and second floor acceleration, respectively, to get the relative value of the acceleration. A numerical integrator is used to compute the velocity and position from the accelerometer signal [21]. The displacement of the floors is calculated using an accelerometer. Since the accelerometer used is a bidirectional one, the acceleration is obtained along X- and Y-components. These accelerations along X- and Y-components are integrated to extract velocity in the first instance and again integrated into extract positions in the next instance, although after each integration, filter is used for the correction of signals. An offset cancellation filter (OCF) is proposed, which completely removes the DC components present in the accelerometer output. In order to avoid the drift caused by low-frequency noise signals, a special high-pass filter is used. A frequency-domain method is used to estimate the low-frequency noise components present in the accelerometer output. The high-pass filter is designed offline according to these noise components. Since the OCF reduces the number of high-pass filtering stages, there is less phase error. The numerical integrator combines the OCF and a high-pass filter.

There is no angular sensor to calculate the angular acceleration of the structure. The angular accelerations are calculated by

$$\ddot{\theta}_1 = -\left(\frac{C_{\theta 01} + C_{\theta 02}}{m_1 r_1^2}\right)\dot{\theta}_1 + \left(\frac{C_{\theta 02}}{m_1 r_1^2}\right)\dot{\theta}_2 - \left(\frac{K_{\theta 01} + K_{\theta 02}}{m_1 r_1^2}\right)\theta_1 + \left(\frac{K_{\theta 02}}{m_1 r_1^2}\right)\theta_2$$
$$+ \left(\frac{C_{x_1} l_{y_1} + C_{x_2} l_{y_2}}{m_1 r_1^2}\right)\dot{x}_1 - \left(\frac{C_{x_2} l_{y_2}}{m_1 r_1^2}\right)\dot{x}_2 + \left(\frac{K_{x_1} l_{y_1} + K_{x_2} l_{y_2}}{m_1 r_1^2}\right)x_1 - \left(\frac{K_{x_2} l_{y_2}}{m_1 r_1^2}\right)x_2$$
$$(3.70)$$
$$- \left(\frac{C_{y_1} l_{x_1} + C_{y_2} l_{x_2}}{m_1 r_1^2}\right)\dot{y}_1 + \left(\frac{C_{y_2} l_{x_2}}{m_1 r_1^2}\right)\dot{y}_2 - \left(\frac{K_{y_1} l_{x_1} + K_{y_2} l_{x_2}}{m_1 r_1^2}\right)y_1 + \left(\frac{K_{y_2} l_{x_2}}{m_1 r_1^2}\right)y_2$$

$$\ddot{\theta}_2 = -\left(\frac{C_{\theta 02}}{m_2 r_2^2}\right)\dot{\theta}_1 - \left(\frac{C_{\theta 02}}{m_2 r_2^2}\right)\dot{\theta}_2 + \left(\frac{K_{\theta 02}}{m_2 r_2^2}\right)\theta_1 - \left(\frac{K_{\theta 02}}{m_2 r_2^2}\right)\theta_2$$
$$+ \left(\frac{C_{x_2} l_{y_2}}{m_2 r_2^2}\right)\dot{x}_1 + \left(\frac{C_{x_2} l_{y_2}}{m_2 r_2^2}\right)\dot{x}_2 - \left(\frac{K_{x_2} l_{y_2}}{m_2 r_2^2}\right)x_1 + \left(\frac{K_{x_2} l_{y_2}}{m_2 r_2^2}\right)x_2 \qquad (3.71)$$
$$+ \left(\frac{C_{y_2} l_{x_2}}{m_2 r_2^2}\right)\dot{y}_1 - \left(\frac{C_{y_2} l_{x_2}}{m_2 r_2^2}\right)\dot{y}_2 - \left(\frac{K_{y_2} l_{x_2}}{m_2 r_2^2}\right)y_1 - \left(\frac{K_{y_2} l_{x_2}}{m_2 r_2^2}\right)y_2,$$

where $\ddot{\theta}_1$ and $\ddot{\theta}_2$ are the angular accelerations of the first and second floors, and C_i, l_i, m_i, and r_i are structural parameters of the building. They are identified by the least square algorithm [22]. The identification process is achieved by identifying the ratio of the parameters corresponding to the mass, damping, and stiffness of a building excited by a seismic activity. The algorithm is based on a parametrization combined with the recursive least square method with forgetting factor. The algorithm is a real time that identifies the parameters of a building model using acceleration measurements of the floors and the ground.

The structural parameters of the two-floor building are identified and fed into the algorithms along with the values of positions and velocities for the calculation of the angular accelerations. The velocities and positions $\dot{x}_1$, $\dot{y}_1$, $\dot{x}_2$, $\dot{y}_2$, and x_1, y_1, x_2, y_2 extracted from the acceleration signals are substituted in the (3.70) and (3.71) to obtain the angular accelerations of ground and top floors. The angular velocities and angular positions $\dot{\theta}_1$, $\dot{\theta}_2$ and θ_1, θ_2 are obtained from angular accelerations $\ddot{\theta}_1$ and $\ddot{\theta}_2$ by using the same numerical integrator.

The theorems of this chapter provide sufficient conditions for the minimal values of the proportional and derivative gains as well as maximum values of the integral gains. For the sake of carrying out a relevant comparison between the PD and PID controllers, it is desirable to use the same proportional and derivative gains.

In this chapter, the PD/PID gains are chosen so as to ensure satisfactory performance as well as within the range specified by the stability theory analysis. The following PD gains are used for the control design:

$$\lambda_m(M_x) = 10, \lambda_m(k_{c_x}) = 20, \lambda_m(k_{c_{x\theta}}) = 8, \lambda_m(M_y) = 10, \lambda_m(k_{c_y}) = 22$$
$$\lambda_m(k_{c_{y\theta}}) = 6, \lambda_{j_t}(J_0) = 5, \lambda_m(k_{c_\theta}) = 21, \lambda_m(k_{c_{x\theta}}) = 8, \lambda_m(k_{c_{y\theta}}) = 6, k_f = 700.$$
(3.72)

From Theorem 3.2, we use the following PID gains:

$$\lambda_m(K_{px}) \geq 1092, \lambda_m(K_{dx}) \geq 55, \lambda_M(K_{ix}) \leq 2324, \lambda_m(K_{py}) \geq 1092$$
$$\lambda_m(K_{dy}) \geq 55, \lambda_M(K_{iy}) \leq 2324, \lambda_{j_0}(K_{p\theta}) \geq 1102, \lambda_{j_0}(K_{d}) \geq 85, \lambda_{J_0}(K_{i\theta}) \leq 3563$$
$$K_{px} = 1800, K_{py} = 2000, K_{p\theta} = 2200, K_{dx} = 160$$
$$K_{dy} = 220, K_{d\theta} = 300, K_{ix} = 2000, K_{iy} = 2300, K_{i\theta} = 3500.$$
(3.73)

Here, the proportional and derivative gains are the same as the PD gains in (3.72).

The performance validation of these controllers is implemented by the vibration control with respect to the seismic execution on the prototype. Northridge earthquake signal is used to vibrate the shake table. The magnitude of 6.7 MW earthquake that occurred near Northridge, California on January 17, 1994 produced an extensive set of strong-motion recordings. The epicenter is located about 32 km nortwest of Los Angeles in the densely populated San Fernando Valley. Analysis by the USGS and Caltech indicates that the earthquake had a thrust mechanism on a fault plane striking N60° W and dipping 35–45° S. The estimated location and magnitude of the Northridge earthquake are as follows:

Epicenter: 34.209° N, 118.541° W. Focal Depth: 19 km (Caltech/USGS). Origin Time: 12:30:55.4, 17 January 1994 UTC (4.30 AM, PST). Magnitude: 6.7 MW (Caltech).

The duration of the strong shaking is about 10–15 s. The peak vertical acceleration is about two-thirds of the peak horizontal. The displacement of the Northridge earthquake is scaled from 16.92 to 1.50 cm, whereas the time is scaled from 39.98 to 11.91 s. This is done to suit the experimental conditions as the maximum allowed movement of the shake table from the reference point on either side is 2 cm. So considering the maximum limit of movement, the time is scaled to match the experimental analysis. The control object is to minimize the relative displacement of each floor in bidirection. The comparisons between the bidirectional PD and PID vibration controllers are carried out considering three cases: (1) without any active control (No Control); (2) with the TA; and (3) with both the AMD and the TA (AMD+TA). The vibration reductions are in three directions: X-, Y-, and θ-directions.

The average vibration displacement is calculated by the mean squared error as

$$MSE = \frac{1}{N} \sum_{k=1}^{N} x(k)^2,$$

where $x(k)$ is the displacement of the floor, and N is the total data number.

The comparison results of the average vibration displacement are shown in Tables 3.1, 3.2, and 3.3. Here, "↓" sign indicates decrease.

Figures 3.4, 3.5, and 3.6 display the action of the PD control to curb the vibration along X-, Y-, and θ-directions. Tables 3.1, 3.2, and 3.3 represent the quantitative analysis of vibration control with both the actuators using PD/PID control along X-, Y-, and θ-directions. If we analyze Figs. 3.4, 3.5, and 3.6 as well as Tables 3.1 and 3.2, it can be observed that AMD performs good in the vibration control along X- and Y-

Table 3.1 Average vibration displacement by AMD+TA

Direction	PD control	% ↓ error	PID control	% ↓ error	No control
X	0.2987	60.2	0.2373	68.5	0.7514
Y	0.0719	45.53	0.0783	59.3	0.1320
θ	0.0696	40.8	0.0611	47.6	0.1174

Table 3.2 Average vibration displacement with PD control

Direction	with AMD	% ↓ error	with TA	% ↓ error	No control
X	0.4832	35.7	0.5802	22.78	0.7514
Y	0.0981	25.68	0.1012	23.3	0.1320
θ	0.0902	23.1	0.0801	31.7	0.1174

Table 3.3 Average vibration displacement with PID control

Direction	with AMD	% ↓ error	with TA	% ↓ error	No control
X	0.3632	51.6	0.4911	34.6	0.7514
Y	0.0849	35.6	0.0969	26.5	0.1320
θ	0.0811	30.8	0.0713	40.0	0.1174

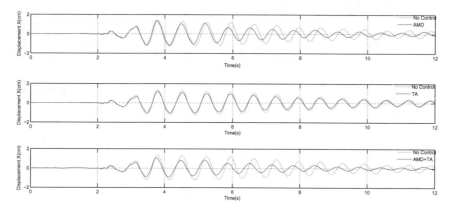

Fig. 3.4 PD control of the second floor in X-direction

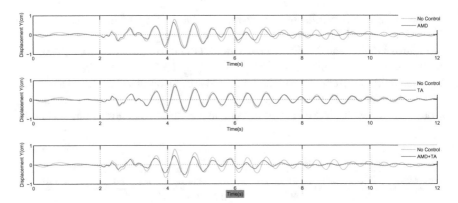

Fig. 3.5 PD control of the second floor in Y-direction

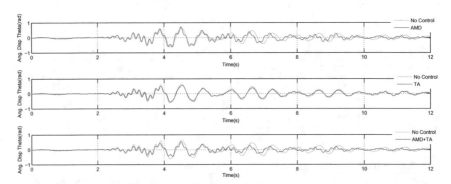

Fig. 3.6 PD control of the second floor in θ-direction

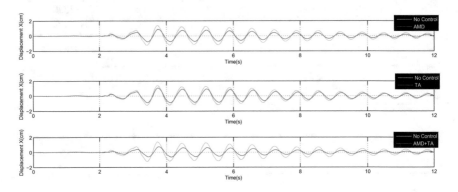

Fig. 3.7 PID control of the second floor in X-direction

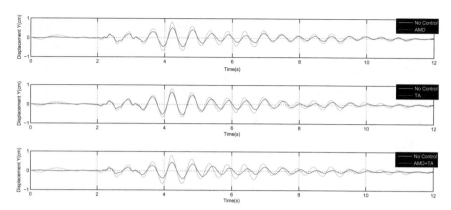

Fig. 3.8 PID control of the second floor in Y-direction

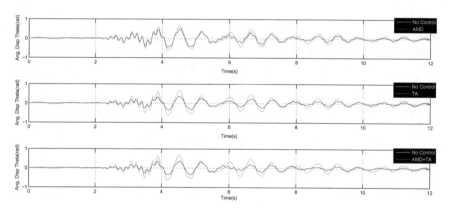

Fig. 3.9 PID control of the second floor in θ-direction

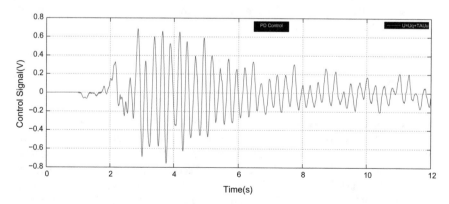

Fig. 3.10 The PD control signal

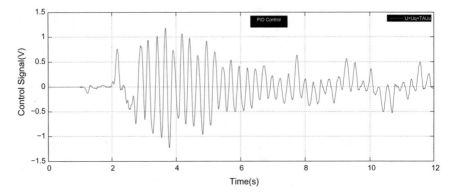

Fig. 3.11 PID control signal

directions but fails to mitigate vibration to a suitable extent along the θ-direction. But if we analyze the behavior of TA, it can be observed that TA performs superiorly in mitigating the vibration along θ-direction but fails to attenuate vibration to acceptable extent along X- and Y-directions. When AMD and TA act simultaneously, it is observed that the results of vibration attenuation are much better along X-, Y-, and θ-directions. The main reason behind this is that superior vibration control is achieved in combination as respective actuators perform well in their assigned zones. Figures 3.7, 3.8, and 3.9 display the action of the PID control to curb the vibration along X-, Y-, and θ-directions. The analysis of PID control from Tables 3.1, 3.2, and 3.3 reveals that PID controller performs much better than PD controller in controlling vibration along X-, Y-, and θ-directions. The behavior of AMD and TA with PID controller follows the same pattern as with the PD controller, and it can be observed from the figures and quantitative analysis that combined action of the actuators achieves high vibration attenuation. Figure 3.10 and Fig. 3.11 represent the control signal of PD control and PID control, respectively.

3.6   Conclusion

In this chapter, the equations of active control of both the actuators (AMD and TA) are proposed. A two-floor structure associated with one AMD and one TA for active vibration control is proposed. The theoretical contribution is the stability analysis for the bidirectional PD/PID control. The sufficient conditions of stability are extracted in order to tune the PD/PID gains. The two theorems stated in this chapter validate the conditions which are sufficient for selecting the minimum values of the proportional and derivative gains. Also, the minimum values of the integral gains are extracted on the basis of Theorem 3.2. The range of the proportional, derivative, and integral gains are specified using the Lyapunov stability theorem and is assured that on the basis of the selected gains, the controller performs in the superior manner. It is observable that both PD and PID controllers work well with AMD and TA. The experimental results show that the PID controller is better than the PD controller in minimizing the vibration in all three directions. By comparing the quantitative analysis as displayed in Tables 3.1, 3.2, and 3.3, it can be concluded that the PID controller in combination with both AMD and TA is considered to be the most efficient in mitigation of vibration along X-, Y-, and θ-directions. The use of TA facilities the torsional vibration attenuation but it fails to attenuate vibration along X- and Y-directions to a considerable extent with PD controller, but when PID controller is used, the vibration control using TA along X- and Y-directions is much better. The vibration attenuations along X- and Y-directions are effectively achieved by AMD but it fails to attenuate vibration considerably along θ-direction with PD controller in comparison with PID controller. As we can see from the quantitative analysis that with PID controller, the mitigation of vibration along θ-direction with AMD is slightly better. Further work on the design of TA is needed so that it can minimize the vibrations to a suitable extent along all three directions.

References

1. W. Park, K.S. Park, H.M. Koh, Active control of large structures using a bilinear pole-shifting transform with H_∞ control method. Eng. Struct. **30**(11), 3336–3344 (2008)
2. H. Du, N. Zhang, H_∞ control for buildings with time delay in control via linear matrix inequalities and genetic algorithms. Eng. Struct. **30**(1), 81–92 (2008)
3. K. Seto, A structural control method of the vibration of flexible buildings in response to large earthquake and strong winds, in *Proceedings of the 35th Conference on Decision and Control,* vol. 1 (1996), pp. 658–663
4. A. Alavinasab, H. Moharrami, A. Khajepour, Active control of structures using energy-based LQR method, in *Computer-Aided Civil and Infrastructure Engineering,* vol. 21, no. 8 (2006), pp. 605–611
5. C.C. Ho, C.K. Ma, Active vibration control of structural systems by a combination of the linear quadratic Gaussian and input estimation approaches. J., Sound Vibr. **301**(3–5), 429–449 (2007)
6. J.C.H. Chang, T.T. Soong, Structural control using active tuned mass damper. J. Eng. Mech. **106**(6), 1091–1098 (1980)
7. A. Yanik, J.P. Pinelli, H. Gutierrez, Control of a three-dimensional structure with magneto-rheological dampers, in *11th International Conference on Vibration Problems,* ed. by Z. Dimitrovová et al., Lisbon, Portugal (2013)
8. R. Guclu, H. Yazici, Vibration control of a structure with ATMD against earthquake using fuzzy logic controllers. J. Sound Vibr. **318**(1–2), 36–49 (2008)
9. S.M. Nigdeli, M.H. Boduroglu, Active tendon control of torsionally irregular structures under near-fault ground motion excitation, in *Computer-Aided Civil and Infrastructure Engineering,* vol. 28, no. 9 (2013), pp. 718–736
10. S.M. Nigdeli, Effect of feedback on PID controlled active structures under earthquake excitations. Earthq. Struct. **6**(2), 217–235 (2014)
11. A.H. Heidari, S. Etedali, M.R. Javaheri-Tafti, A hybrid LQR-PID control design for seismic control of buildings equipped with ATMD, in *Frontiers of Structural and Civil Engineering,* vol. 12, no. 44 (2018)

12. C. Roldán, F.J. Campa, O. Altuzarra et al., Automatic identification of the inertia and fric-
    tion of an electromechanical actuator, in *New Advances in Mechanisms, Transmissions and
    Applications*. Volume 17 of the series Mechanisms and Machine Science (2014), pp. 409–416
13. F.L. Lewis, D.M. Dawson, C.T. Abdallah, *Robot Manipulator Control: Theory and Practice*,
    2nd edn. (Marcel Dekker Inc., 2004)
14. K. De Cock, B. De Moor, W. Minten et al., A tutorial on PID-control. Katholieke Universiteit
    Leuren Department of Electrical Engineering ESAT-SISTA /TR (1997)
15. R. Saragih, Designing active vibration control with minimum order for flexible structures, in
    IEEE International Conference on Control and Automation (2012), pp. 450–453
16. E.D. Sontag, Y. Wang, On characterizations of the input-to-state stability property, Syst. Control
    Lett. **24**(5), 351–359 (1995)
17. Shake Table I-40, *User Manual, STI-40* (Quanser Inc., 2012)
18. Models STB1104-1116 Servotube Component, *User Manual,* Copley Motion Systems LLC
19. IINTECO RT-DAC/USB2 I/O Board, *User's Manual,* Kraków (2010)
20. XL403A Acceleroemeter, *User Manual,* (Measurement Specialities, 2013)
21. S. Thenozhi, W. Yu, Advances in modeling and vibration control of building structures. Ann.
    Rev. Control **37**(2), 346–364 (2013)
22. S. Thenozhi,, W. Yu, Stability analysis of active vibration control of building structures using
    PD/PID control. Eng. Struct. **81**(7), 208–218 (2014)

Chapter 4
Type-2 Fuzzy PD/PID Control
of Structures

4.1 Introduction

Fuzzy logic has obtained many attentions in control of structure because of its simple nature, robustness, and nonlinear mapping capability (Choi et al., Int J Solids Struct 42(16–17):4779–4794, [1]). Reigles and Symans (Structural control and health monitoring, pp 724–747, [2]) present a numerical study to show the effectiveness of a supervisory fuzzy logic controller for seismic response control of an eight-storey base-isolated structure affected by translation–torsional motion. The combination of fuzzy logic and PD control has more degrees of freedom to tune the mass damper for the structure vibration [3]. In [4], a fuzzy supervisory method is used for the active control of building structures. In [5], the semi-active control of building under the earthquake is implemented with the concept of fuzzification related to magnetorheological (MR) damper characteristics. The concept of optimal fuzzy control is utilized to the structure under the seismic forces in [6]. An optimal fuzzy control for suppressing vibration of buildings by utilizing MR dampers is proposed in [7]. Fuzzy control with hybrid mass damper for the torsionally couple problem is discussed in [8]. The implementation of explicit formula of hedge-algebras-based fuzzy controller in the control of structural vibration was demonstrated in [9]. The concept of type-2 fuzzy sets has been presented in [10]. The type-2 fuzzy system has effective ways to deal with knowledge uncertainty compared with classical type-1 fuzzy logic, because the type-2 fuzzy sets can deal uncertainties with more parameters and more design degrees of freedom [11]. A simplified type-2 fuzzy system can be applied in the real-time application [12]. For the vibration control of single degree of freedom, [13] uses active tuned mass damper and type-2 fuzzy control. In [14], a semi-active tuned mass damper combined with adaptive MR damper is utilized, and the type-1 and type-2 fuzzy controllers are implemented.

In this chapter, we will use the type-2 fuzzy control to compensate the regulation errors of PD and PID controllers. The main parts of the controllers are PD and PID, while the nonlinearity is compensated by the type-2 fuzzy system. We use a disk with

motor arrangement as the torsional actuator (TA) to minimize the torsional response
of the building. The sufficient conditions for asymptotic stability of PD/PID with
type-2 fuzzy system are validated. The conditions are quite convenient for the design
of the controller gains. An active vibration control system is designed with two-floor
building structure equipped with active mass damper (AMD) and TA. The experimental results are obtained by using the type-2 fuzzy PD/PID controllers. Compared
to the other active vibration controllers, our type-2 Fuzzy PD/PID controllers do not
have big derivative and integral gains. So the results of our active controllers are
much better than the others.

4.2   PD Control with Type-2 Fuzzy Compensation

The AMD and TA are placed on the structure in a similar manner as mentioned in
Chap. 3. All the dynamic equations related to AMD and TA are also mentioned in
Chap. 2. The closed-loop system (3.1) with the control $\mathbf{u}$ is

$$M\ddot{x} + C\dot{x} + Sx + f_e + \Gamma d_{\mathbf{u}} = \Gamma(-K_p x - K_d \dot{x}). \qquad (4.1)$$

The active vibration control of the building structure can be regarded as the regulation problem with zero reference,

$$u = -K_p x - K_d \dot{x}, \qquad (4.2)$$

where $x = [x, y, \theta]^T$. Here the reference $x^d = \dot{x}^d = 0$, the regulation error $e = x - x^d = x$, K_p, and K_d are positive-definite constant matrices that correspond to the
proportional and derivative gains. The closed-loop system with the PD control is

$$M\ddot{x}(t) + C\dot{x}(t) + F = -\Gamma\left(K_p \mathbf{x} + K_d \dot{\mathbf{x}}\right), \qquad (4.3)$$

where F is the uncertainty,

$$F = Sx + f_e + \Gamma d_{\mathbf{u}} = [f_x, f_y, f_\theta]^T, \qquad (4.4)$$

Since F is unknown, we use a fuzzy system to approximate it. Compared with
normal fuzzy sets, type-2 fuzzy sets can model easily for big magnitude uncertainties
with less fuzzy rules. The membership functions of type-2 fuzzy systems are not
longer the crisp values; they are in the interval of [0, 1] [10, 11].
    A generic fuzzy model for the uncertainty F in ith floor is provided by a collection
of p fuzzy rules for x, y, and θ:

R^i: IF $(x_i$ is $A_{1i})$ and $(y_i$ is $A_{2i})$ and $(\theta_i$ is $A_{3i})$ and $(\dot{x}_i$ is $A_{4i})$ and $(\dot{y}_i$ is $A_{5i})$
    and $\left(\dot{\theta}_i$ is $A_{6i}\right)$ THEN $(f_x$ is $B_{1i})$ and $\left(f_y$ is $B_{2i}\right)$ and $(f_\theta$ is $B_{3i})$,

$$(4.5)$$

where $A_{1i} \ldots, A_{6i}, B_{1i}, B_{3i}$, and B_{3i} are type-2 fuzzy sets. The type-2 fuzzy set A with the membership function G_A is defined as

$$A = \{(x, \varsigma), G_A(x, \varsigma) \mid \forall x \epsilon R, \forall \varsigma \epsilon M_x \subseteq [0, 1]\}, \tag{4.6}$$

where ς is an auxiliary variable, and $0 \leqslant G_A(x, \varsigma) \leqslant 1$, M_x is the primary membership function. For the type-2 fuzzy set A,

$$A = \int_{x \epsilon X} \int_{\varsigma \epsilon M_x} G_A(x, \varsigma)/(x, \varsigma).$$

The integral $\int$ of the classical fuzzy set becomes the sum $\sum$.

The upper and lower membership functions are defined as $G_A^u(x_1, \varsigma)$ and $G_A^l(x, \varsigma)$. They describe the upper and lower bounds of the uncertainties. For ith rule and the point x_1, the crisp input is fuzzified in the interval of $[f_i^l(x_1), f_i^u(x_1)]$,

$$\begin{aligned} f_i^u(x_1) &= G_{A_{1i}}^u(x_1, \varsigma) * G_{A_{2i}}^u(x_1, \varsigma) * G_{A_{3i}}^u(x_1, \varsigma) \\ f_i^l(x_1) &= G_{A_{1i}}^l(x_1, \varsigma) * G_{A_{2i}}^l(x_1, \varsigma) * G_{A_{3i}}^l(x_1, \varsigma), \end{aligned} \tag{4.7}$$

where $*$ denote t-norm operator; it can be the minimization. For all l rules, type-2 fuzzy inference engine aggregates with the fuzzified inputs and infers another type-2 fuzzy set,

$$G_O(y) = \sqcup_{x \epsilon X}[G_A(x) \sqcap G_B(x, y)]. \tag{4.8}$$

We use the type-reduction method to convert $G_O(y)$ into type-1 fuzzy set. This technique captures more information about rule uncertainties than does the defuzzified value (a crisp number) and seems to be as fundamental to the design of fuzzy logic systems that include linguistic uncertainties (that translate into rule uncertainties) as variance is to the mean in case of probabilistic uncertainties. The centroids associated with type-2 fuzzy sets are calculated. For ith rule, the centroid of jth output fuzzy rule (4.5) is $y_j^i = [y_{lj}^i, y_{rj}^i]$, and y_{lj}^i and y_{rj}^i are the most left and right points. The type-2 fuzzy sets are reduced to the type-1 fuzzy set with the interval $[y_{lj}^i, y_{rj}^i]$. The most popular technique for type-reducing an interval type-2 fuzzy set is the Karnik–Mendel (KM) iterative procedure [15]. The outcome of type-reduction of an interval type-2 fuzzy set is an interval type-1 set considering the criteria that the centroid is placed between the two endpoints. The iterative methodology is a superior technique in order to find these endpoints. The centroid of the type-1 set is considered to be the center of this interval. Mendel [16] laid down the main design criteria in consideration to type-2 uncertainty measurement as when all sources of uncertainty disappear, a type-2 fuzzy logic system must reduce to a comparable type-1 fuzzy logic system. This proposed statement is valid throughout. So it is valid from the proposed statement that is no uncertainty associated with a type-1 fuzzy set, and therefore the mentioned measures of uncertainty for type-1 fuzzy sets [17] cannot be measuring uncertainty; instead, they are measuring separate aspect of the type-1 set to be mentioned as vagueness. For all p rules,

$$y_{lj} = \frac{\sum_{i=1}^{p} f_l^i y_{lj}^i}{\sum_{i=1}^{p} f_l^i}, \quad y_{rj} = \frac{\sum_{i=1}^{p} f_r^i y_{rj}^i}{\sum_{i=1}^{p} f_r^i}, \tag{4.9}$$

where f_l^i and f_r^i are the firing strengths associated with y_{lj}^i and y_{rj}^i of ith rule. By the minimization and maximization operations, y_{lj} and y_{rj} can be expressed as

$$y_{lj} = \frac{\sum_{i=1}^{p} f_{lj}^i y_{lj} + \sum_{i=1}^{p} f_{rj}^i y_{lk}}{\sum_{i=1}^{q} f_r^i + \sum_{i=1}^{q} f_l^i}, \quad y_{rj} = \frac{\sum_{i=1}^{p} f_{lj}^i y_{rj} + \sum_{i=1}^{p} f_{rj}^i y_{rk}}{\sum_{i=1}^{q} f_r^i + \sum_{i=1}^{q} f_l^i}, \tag{4.10}$$

where $q_{lj}^i = \frac{f_l^i}{\sum_{i=1}^{q} f_r^i + \sum_{i=1}^{q} f_l^i}$, $q_{rj}^i = \frac{f_r^i}{\sum_{i=1}^{q} f_r^i + \sum_{i=1}^{q} f_l^i}$. By singleton fuzzifier, the jth output of the fuzzy logic system can be expressed as

$$\hat{f}_j = \frac{y_{rj} + y_{lj}}{2} = \frac{1}{2} \left[(\phi_{rj}^T(z) w_{rj}(z) + \phi_l^T(z) w_{lj}(z) \right], \tag{4.11}$$

where $j = 1, 2, 3$. w_{rj} is the point at which $\mu_{B_{rj}} = 1$, w_{lj} is the point at which $\mu_{B_{lj}} = 1$, $z = \left[x, y, \theta, \dot{x}, \dot{y}, \dot{\theta} \right]^T$. In matrix form, the estimation of the uncertainty $\mathbf{F}$ is

$$\hat{\mathbf{F}} = \frac{1}{2} \left[\mathbf{\Phi}_r^T(z) W_r(z) + \mathbf{\Phi}_l^T(z) W_l(z) \right], \tag{4.12}$$

where $\hat{\mathbf{F}} = \left[\hat{f}_1, \hat{f}_2, \hat{f}_3 \right] = [\hat{f}_x, \hat{f}_y, \hat{f}_\theta]^T$.

PD control with type-2 fuzzy compensation is

$$\mathbf{u} = -K_p \mathbf{x} - K_d \dot{\mathbf{x}} - \frac{1}{2} \mathbf{\Phi}_r^T(z) W_r(z) - \frac{1}{2} \mathbf{\Phi}_l^T(z) W_l(z). \tag{4.13}$$

The new closed-loop system is

$$\begin{bmatrix} M_x \ddot{x} \\ M_y \ddot{y} \\ J_t \ddot{\theta} \end{bmatrix} + \begin{bmatrix} C_x \dot{x} - C_{x\theta} \dot{\theta} \\ C_y \dot{y} + C_{y\theta} \dot{\theta} \\ C_\theta \dot{\theta} - C_{x\theta} \dot{x} + C_{y\theta} \dot{y} \end{bmatrix} + \begin{bmatrix} f_x \\ f_y \\ f_\theta \end{bmatrix}$$
$$= -\Gamma \begin{bmatrix} K_{px} x \\ K_{py} y \\ K_{p\theta} \theta \end{bmatrix} - \Gamma \begin{bmatrix} K_{dx} \dot{x} \\ K_{dy} \dot{y} \\ K_{d\theta} \dot{\theta} \end{bmatrix} - \frac{1}{2} \Gamma \begin{bmatrix} \phi_r^T(z_x) w_r(z_x) + \phi_l^T(z_x) w_l(z_x) \\ \phi_r^T(z_y) w_r(z_y) + \phi_l^T(z_y) w_l(z_y) \\ \phi_r^T(z_\theta) w_r(z_\theta) + \phi_l^T(z_\theta) w_l(z_\theta) \end{bmatrix},$$
$$\tag{4.14}$$

where $K_{px}, K_{dx}, K_{py}, K_{dy}, K_{p\theta}, K_{d\theta}$ are the gains considering X-, Y-, and θ-components, respectively. Because the three components, x, y, and θ, have the same form, in the following stability analysis, we only discuss X-component. Y- and θ-components have similar results. In order to simplify the controller, we let $\Gamma = I$.

Because $K_{px}x + K_{dx}\dot{x} = K_x(\Lambda x + \dot{x})$, Λ is the positive-definite matrix, $K_x \Lambda = K_{px}$, $K_x = K_{dx}$, we define an auxiliary variable r_x as

$$r_x = \dot{x} + \Lambda x. \tag{4.15}$$

So

$$u_x = -K_x r_x - \frac{1}{2}\phi_r^T(z_x)w_r(z_x) - \frac{1}{2}\phi_l^T(z_x)w_l(z_x), \tag{4.16}$$

where $z_x = (x, \dot{x})$. The closed-loop system of X-component becomes

$$M_x\ddot{x} + C_x\dot{x} - C_{x\theta}\dot{\theta} + f_x = u_x. \tag{4.17}$$

The X-component also gives the torsion in the structure

$$\dot{\theta} = -\alpha_f\dot{x}. \tag{4.18}$$

Equation (4.17) becomes

$$M_x\ddot{x} + C_x\dot{x} + \alpha_f C_{x\theta}\dot{x} + f_{\mathbf{x}} = u_x. \tag{4.19}$$

Because $M_x\dot{r}_x = M_x(\ddot{x} + \Lambda\dot{x})$, using (4.19) we have

$$M_x\dot{r}_x + C_x r_x + \alpha_f C_{x\theta} r_x = u_x + \Delta f_x, \tag{4.20}$$

where $\Delta f_x = M_x\Lambda\dot{x} + C_x\Lambda x + \alpha_f C_{x\theta}\Lambda x - f_{\mathbf{x}}$.

With Stone–Weierstrass theorem [18], Δf_x can be estimated by the type-2 fuzzy system (4.12) as

$$\Delta f_x = \frac{1}{2}\phi_r^T(z_x)w_r^*(z_x) + \frac{1}{2}\phi_l^T(z_x)w_l^*(z_x) + \sigma_x, \tag{4.21}$$

where σ_x is the modeling error, and w_r^* and w_l^* are unknown optimal weights. We assumed that it is bounded as

$$\sigma_x^T\Lambda_\sigma^{-1}\sigma_x \leq \bar{\sigma}_x, \tag{4.22}$$

where Λ_σ is a known positive-definite matrix.

With the type-2 fuzzy PD control (4.16), the closed-loop system (4.20) becomes

$$M_x\dot{r}_x + C_x r_x + \alpha_f C_{x\theta} r_x = -K_x r_x - \frac{1}{2}\phi_r^T(z_x)\tilde{w}_r(z_x) - \frac{1}{2}\phi_l^T(z_x)\tilde{w}_l(z_x) + \sigma_x, \tag{4.23}$$

where $\tilde{w}_r = w_r - w_r^*$, $\tilde{w}_l = w_l - w_l^*$.

The following theorem gives the stability analysis of type-2 fuzzy PD control (4.13) with a gradient descent algorithm for $w_r(z_x)$ and $w_l(z_x)$. The major advantage of this method is that fuzzy rules or membership functions can be learned without changing the form of the fuzzy rule table used in usual fuzzy controls, so that the case of weak-firing can be avoided well, which is different from the conventional learning algorithm.

Theorem 4.1 *Consider the structural system (4.1) controlled by the type-2 fuzzy PD controller as (4.16), if the gain satisfies*

$$K_x > \Lambda_\sigma. \tag{4.24}$$

Λ_σ *is defined in (4.22), and the fuzzy system is updated as*

$$\frac{d}{dt} w_r(z_x) = -[k_w r_x^T \phi_r^T(z_x)]^T \\ \frac{d}{dt} w_l(z_x) = -[k_w r_x^T \phi_l^T(z_x)]^T \tag{4.25}$$

$k_w > 0$, *then the filter regulation errors* r_x *and* r_θ *converge to the residual sets*

$$D_x = \left\{ r_x | \, \|r_x\|^2 \leq \bar{\sigma}_x \right\} \\ D_\theta = \left\{ r_\theta | \, \|r_\theta\|^2 \leq \alpha_f^2 \bar{\sigma}_x \right\}. \tag{4.26}$$

$\bar{\sigma}_x$ *is defined in (4.22).*

Proof Since M_x and Λ are positive-definite matrices, let us consider Lyapunov candidate V_x for the X-component

$$V_x = \frac{1}{2} r_x^T M_x r_x + \frac{1}{4} tr_x[\tilde{w}_r^T(z_x) k_w^{-1} \tilde{w}_r(z_x)] + \frac{1}{4} tr_x[\tilde{w}_l^T(z_x) k_w^{-1} \tilde{w}_l(z_x)]. \tag{4.27}$$

Using (4.23) and $r_x^T (\dot{M}_x - 2C_x) r_x = 0$, the derivative of (4.27) is

$$\dot{V}_x = -r_x^T K_x r_x - \frac{1}{2} r_x^T \phi_r^T(z_x) \tilde{w}_r(z_x) - \frac{1}{2} r_x^T \phi_l^T(z_x) \tilde{w}_l(z_x) + r_x^T \sigma_x - r_x^T \alpha_f C_{x\theta} r_x \\ + \frac{1}{2} tr_x[\tilde{w}_r^T(z_x) k_w^{-1} \frac{d}{dt} \tilde{w}_r(z_x)] + \frac{1}{2} tr_x[\tilde{w}_l^T(z_x) k_w^{-1} \frac{d}{dt} \tilde{w}_l(z_x)]. \tag{4.28}$$

Using the updating law (4.25), and

$$\frac{1}{2} tr_x[\tilde{w}_r^T(z_x) k_w^{-1} \frac{d}{dt} \tilde{w}_r(z_x)] - \frac{1}{2} r_x^T \phi_l^T(z_x) \tilde{w}_l(z_x) = 0 \\ \frac{1}{2} tr_x[\tilde{w}_l^T(z_x) k_w^{-1} \frac{d}{dt} \tilde{w}_l(z_x)] - \frac{1}{2} r_x^T \phi_l^T(z_x) \tilde{w}_l(z_x) = 0,$$

Equation (4.28) becomes

$$\dot{V}_x = -r_x^T K_x r_x + r_x^T \sigma_x - r_x^T \alpha_f C_{x\theta} r_x. \tag{4.29}$$

Now let us consider matrix inequality as follows:

$$X^T Y + Y^T X \leq X^T \Lambda X + Y^T \Lambda^{-1} Y \tag{4.30}$$

for all $X, Y \in R^n$, $0 < \Lambda = \Lambda^T$. From (4.22), $r_x^T \sigma_x$ can be estimated as

$$r_x^T \sigma_x \leq r_x^T \Lambda_\sigma r_x + \sigma_x^T \Lambda_\sigma^{-1} \sigma_x \leq r_x^T \Lambda_\sigma r_x + \bar{\sigma}_x, \tag{4.31}$$

where $\Lambda_\sigma > 0$. Since $\alpha_f C_{x\theta} > 0$, (4.29) is

$$\dot{V}_x \leq -r_x^T (K_x + \alpha_f C_{x\theta} - \Lambda_\sigma) r_x + \sigma_x^T \Lambda_\sigma^{-1} \sigma_x$$
$$\leq -r_x^T (K_x - \Lambda_\sigma) r_x + \sigma_x^T \Lambda_\sigma^{-1} \sigma_x. \tag{4.32}$$

We can choose the gain of the PD control (4.16), such that (4.24) is established, then

$$\dot{V}_x \leq -\|r_x\|_{K_1}^2 + \bar{\sigma}_x, \tag{4.33}$$

where $K_1 = K_x - \Lambda_\sigma \cdot V_x$ is considered to be ISS-Lyapunov function. In this case, $r_x = \dot{x} + \Lambda x$ is bounded when σ_x is bounded by $\bar{\sigma}_x$ [19]. Because $\int_0^T \dot{V}_x = V_T - V_0 \leq -\int_0^T r_x^T K_x r_x dt + \bar{\sigma}_x T$,

$$\lim_{T \to \infty} \frac{1}{T} \int_0^T \|r_x\|_{K_1}^2 dt \leq \bar{\sigma}_x.$$

Since $\dot{x} = -\alpha_f^{-1} \dot{\theta}$ and $|r_x| = -\alpha_f^{-1} |r_\theta|$, the filter regulation error for θ can be

$$r_\theta = \dot{\theta} + \Lambda \theta. \tag{4.34}$$

So

$$\dot{V}_x \leq -\|r_\theta\|_{K_2}^2 + \bar{\sigma}_x, \tag{4.35}$$

where $K_2 = \alpha_f^{-2} K_1$. So $r_\theta = \dot{\theta} + \Lambda \theta$ is bounded, and

$$\lim_{T \to \infty} \frac{1}{T} \int_0^T \|r_\theta\|_{K_1}^2 dt \leq \alpha_f^2 \bar{\sigma}_x. \tag{4.36}$$

They are (4.26).

Remark 4.1 Compared with the fuzzy compensation (4.13), the advantage of adaptive fuzzy compensation (4.23) is that we do not need to be concerned about the big compensation error in equation (4.21), which results from a poor membership function selection. The gradient algorithms (4.25) ensure that the membership functions $w_r(z_x)$ and $w_l(z_x)$ are updated such that the regulation errors r_x and r_θ are reduced. The above theorem also guarantees that the updating algorithms are stable. It is well known that the regulation error becomes smaller while increasing the derivative gain. The cost of large derivative gain results in slow transient performance. Only when derivative gain tends to infinity, the regulation error converges to zero [20]. However, it would seem better to use a smaller derivative gain if the system contains high-frequency noise signals. In order to decrease the steady-state errors caused by these uncertainties, the derivative gain K_d has to be increased. The transient performances are worsened, for example, the response becomes slow.

4.3   PID Control with Type-2 Fuzzy Compensation

The utilization of fuzzy compensation results in the decrease of regulation error as mentioned in Theorem 4.1. Considering the control theory, the steady-state error can be removed to more extent by introducing an integral component to the PD control. PID controllers use feedback strategy and have three actions: P action is introduced for increasing the speed of response; D action is introduced for damping purposes; and I action is introduced for obtaining a desired steady-state response. Considering the effect of all three components of control, the control law is PID control

$$
\mathbf{u} = -K_p \mathbf{x} - K_i \int_0^t \mathbf{x} d\tau - K_d \dot{\mathbf{x}}, \tag{4.37}
$$

where K_p, K_i, and K_d are positive-definite, and K_i is the integration gain, for the structure control considering the reference $x^d = \dot{x}^d = 0$.

A big integration gain causes unacceptable transient performances and stability problems. Same as of type-2 fuzzy PD control, a type-2 fuzzy compensator for PID control can be applied. PID control with type-2 fuzzy control is

$$
\mathbf{u} = -K_p \mathbf{x} - K_i \int_0^t \mathbf{x} d\tau - K_d \dot{\mathbf{x}} - \frac{1}{2}\boldsymbol{\Phi}_r^T(z)\hat{W}_r(z) - \frac{1}{2}\boldsymbol{\Phi}_l^T(z)\hat{W}_l(z). \tag{4.38}
$$

Similar with PD control, we only consider the X-component (4.19),

$$
u_x = -K_{px}x - K_{dx}\dot{x} - K_{ix}\int_0^t x d\tau - \frac{1}{2}\phi_r^T(z_x)w_r(z_x) - \frac{1}{2}\phi_l^T(z_x)w_l(z_x). \tag{4.39}
$$

In matrix form, (4.19) is

$$
\frac{d}{dt}\begin{bmatrix} s_x \\ x \\ \dot{x} \end{bmatrix} = \begin{bmatrix} K_{ix}x \\ \dot{x} \\ -M_x(C_x\dot{x} + \alpha_f C_{x\theta}\dot{x} + f_x - u_x) \end{bmatrix}, \tag{4.40}
$$

where s_x is auxiliary variable. By using (4.21), the closed-loop system is

$$M_x\ddot{x} + \left(C_x + \alpha_f C_{x\theta}\right)\dot{x} + \tfrac{1}{2}\phi_r^T(z_x)w_r^*(z_x) + \tfrac{1}{2}\phi_l^T(z_x)w_l^*(z_x) + \sigma_x$$
$$= -K_{px}x - K_{dx}\dot{x} - s_x - \tfrac{1}{2}\phi_r^T(z_x)\tilde{w}_r(z_x) - \tfrac{1}{2}\phi_l^T(z_x)\tilde{w}_l(z_x). \tag{4.41}$$

The equilibrium of (4.40) is $[s_x, x, \dot{x}] = [s_x^*, 0, 0]$. Since at equilibrium point $x = 0$ and $\dot{x} = 0$, the equilibrium is $[\sigma_x(0), 0, 0]$. In order to move the equilibrium to origin, we define

$$\tilde{s}_x = s_x - \sigma_x(0), \tag{4.42}$$

where σ_x is the unknown modeling error. The closed-loop system becomes

$$M_x\ddot{x} + \left(C_x + \alpha_f C_{x\theta}\right)\dot{x} + \tfrac{1}{2}\phi_r^T(z_x)w_r^*(z_x) + \tfrac{1}{2}\phi_l^T(z_x)w_l^*(z_x) + \sigma_x$$
$$= -K_{px}x - K_{dx}\dot{x} - \tilde{s}_x - \tfrac{1}{2}\phi_r^T(z_x)\tilde{w}_r(z_x) - \tfrac{1}{2}\phi_l^T(z_x)\tilde{w}_l(z_x) + \sigma_x(0) \tag{4.43}$$
$$\tfrac{d}{dt}\tilde{s}_x = K_{ix}x,$$

where $\tilde{w}_r = w_r - w_r^*$, $\tilde{w}_l = w_l - w_l^*$.

We need the following properties to prove stability of fuzzy PID control.

P1. The positive-definite matrix M_x satisfies the following condition:

$$0 < \lambda_m(M_x) \leq \|M_x\| \leq \lambda_M(M_x) \leq \bar{m}, \tag{4.44}$$

where $\lambda_m(M_x)$ and $\lambda_M(M_x)$ are the minimum and maximum eigenvalues of the matrix M_x, respectively, and $\bar{m} > 0$ is the upper bound.

P2. The modeling error σ_x is Lipschitz over x_1 and x_2

$$\|\sigma_x(x_1) - \sigma_x(x_2)\| \leq k_{\sigma_x}\|x_1 - x_2\|, \tag{4.45}$$

where k_{σ_x} is the Lipschitz constant. Most of the uncertainties are first-order continuous functions. Since f_{sx}, f_{xe}, and d_{ux} are first-order continuous functions and satisfy Lipschitz condition, **P2** can be established.

Now, we calculate the lower bound of the modeling error $\int \sigma_x dx$,

$$\int_0^t \sigma_x dx = \int_0^t f_{sx}dx + \int_0^t f_{xe}dx + \int_0^t d_{ux}dx$$
$$- \frac{1}{2}\int_0^t \phi_r^T(z_x)w_r(z_x)dx - \frac{1}{2}\int_0^t \phi_l^T(z_x)w_l(z_x). \tag{4.46}$$

Here we define the lower bound of $\int_0^t f_{sx}dx$ as $\bar{f}_{sx}$, $\int_0^t f_{xe}dx$ as $\bar{f}_{xe}$ and $\int_0^t d_{ux}dx$ as $\bar{d}_{ux}$. Compared with f_{sx} and d_{ux}, f_{xe} is much bigger in the case of earthquake. Since $\phi_r^T(z_x)$ and $\phi_l^T(z_x)$ are Gaussian functions,

$$\frac{1}{2} \int_0^t \phi_r^T (z_x) w_r(z_x) dx = \frac{W_r(z)}{4} \sqrt{\pi} F(z)$$

$$\frac{1}{2} \int_0^t \phi_l^T (z_x) w_l(z_x) = \frac{W_l(z)}{4} \sqrt{\pi} F(z).$$

$F(z)$ is an "erf" function.

$$k_{\text{œx}} = -\bar{f}_{sx} - \bar{f}_{xe} - \bar{d}_{ux} - \frac{W_r(z)}{4} \sqrt{\pi} - \frac{W_l(z)}{4} \sqrt{\pi}. \qquad (4.47)$$

Here $\sigma_x (0)$ is considered to be zero as it is concerned to structures. The following theorem gives the stability analysis of type-2 fuzzy PID controller (4.39).

Theorem 4.2 *Consider the structural system as (4.1) controlled by the type-2 fuzzy PID controller as (4.39); the closed-loop system (4.41) is asymptotically stable at the equilibriums*

$$[s_x - \sigma_x (0), x, \dot{x}]^T = 0$$

if the PID control gains satisfy

$$\lambda_m (K_{px}) \geq \frac{3}{2} [k_{\text{œx}} + \lambda_M(C_x) + \lambda_M(\alpha_f C_{x\theta})]$$

$$\lambda_M(K_{ix}) \leq \frac{1}{6} \sqrt{\frac{1}{3} \lambda_m (M_x) \lambda_m (K_{px})} \frac{\lambda_m(K_{px})}{\lambda_M (M_x)} \qquad (4.48)$$

$$\lambda_m (K_{dx}) \geq \frac{1}{4} \sqrt{\frac{1}{3} \lambda_m (M_x) \lambda_m (K_{px})} \left[1 + \frac{\lambda_M(C_x) + \lambda_M(\alpha_f C_{x\theta})}{\lambda_M (M_x)} \right]$$
$$- \lambda_m(C_x) - \lambda_m(\alpha_f C_{x\theta}),$$

where k_w is positive-definite matrix, $\mu_x > 0$ is a design parameter, and $\lambda_m(M)$ and $\lambda_M(M)$ are the minimum and maximum eigenvalues of the matrix M. The updating law for the type-2 fuzzy compensator is

$$\frac{d}{dt} w_r(z_x) = -\left[k_w \left(\dot{x} + \frac{\mu_x}{2} x \right)^T \phi_r^T (z_x) \right]^T$$

$$\frac{d}{dt} w_l(z_x) = -\left[k_w \left(\dot{x} + \frac{\mu_x}{2} x \right)^T \phi_l^T (z_x) \right]^T . \qquad (4.49)$$

Proof Here, the Lyapunov candidate is defined as

$$V = \frac{1}{2}\dot{x}^T M_x \dot{x} + \frac{1}{2}x^T K_{px}x + \frac{\mu_x}{4}\hat{\xi}_x^T K_{ix}^{-1}\hat{\xi}_x + x^T \hat{\xi}_x + \frac{\mu_x}{2}x^T M_x \dot{x} + \frac{\mu_x}{4}x^T K_{dx}x$$
$$+ \int_0^t \sigma_x dx - k_{\mathfrak{e}_x} + \frac{1}{4}tr_x[\tilde{w}_r^T(z_x)k_w^{-1}\tilde{w}_r(z_x)] + \frac{1}{4}tr_x[\tilde{w}_l^T(z_x)k_w^{-1}\tilde{w}_l(z_x)],$$

$$(4.50)$$

where $V(0) = 0$. In order to show that $V \geq 0$, V is separated into three parts, such that $V = V_1 + V_2 + V_3$

$$V_1 = \frac{1}{6}x^T K_{px}x + \frac{\mu_x}{4}x^T K_{dx}x + \int_0^t \sigma_x dx - k_{\mathfrak{e}_x}$$
$$+ \frac{1}{4}tr_x[\tilde{w}_r^T(z_x)k_w^{-1}\tilde{w}_r(z_x)] + \frac{1}{4}tr_x[\tilde{w}_l^T(z_x)k_w^{-1}\tilde{w}_l(z_x)] \geq 0,$$
$$K_{px} > 0, K_{dx} > 0$$

$$(4.51)$$

$$V_2 = \frac{1}{6}x^T K_{px}x + \frac{\mu_x}{4}\hat{\xi}_x^T K_{ix}^{-1}\hat{\xi}_x + x^T \hat{\xi}_x$$
$$\geq \frac{1}{2}\frac{1}{3}\lambda_m(K_{px})\|x\|^2 + \frac{\mu_x \lambda_m(K_{ix}^{-1})}{4}\|s_x\|^2 - \|x\|\|s_x\|. \qquad (4.52)$$

When $\mu_x \geq \frac{3}{(\lambda_m(K_{ix}^{-1})\lambda_m(K_{px}))}$,

$$V_2 \geq \frac{1}{2}\left(\sqrt{\frac{\lambda_m(K_{px})}{3}}\|x\| - \sqrt{\frac{3}{4(\lambda_m(K_{px}))}}\|s_x\|\right)^2 \geq 0 \qquad (4.53)$$

and

$$V_{3x} = \frac{1}{6}x^T K_{px}x + \frac{1}{2}\dot{x}^T M\dot{x} + \frac{\mu_x}{2}x^T M\dot{x}. \qquad (4.54)$$

Because

$$Y^T AX \geq \|Y\|\|AX\| \geq \|Y\|\|A\|\|X\| \geq \lambda_M(A)\|Y\|\|X\| \qquad (4.55)$$

when

$$\mu_x \leq \frac{1}{2}\frac{\sqrt{\frac{1}{3}\lambda_m(M_x)\lambda_m(K_{px})}}{\lambda_M(M_x)}$$

$$V_3 \geq \frac{1}{2}\left(\frac{1}{3}\lambda_m(K_{px})\|x\|^2 + \lambda_m(M_x)\|\dot{x}\|^2 + 2\mu_x\lambda_M(M_x)\|x\|\|\dot{x}\|\right)$$
$$= \frac{1}{2}\left(\sqrt{\frac{\lambda_m(K_{px})}{3}}\|x\| + \sqrt{\lambda_m(M_x)}\|\dot{x}\|\right)^2 \geq 0. \qquad (4.56)$$

Now, we have

$$\frac{1}{2}\frac{\sqrt{\frac{1}{3}\lambda_m(M_x)\lambda_m(K_{px})}}{\lambda_M(M_x)} \geq \mu_x \geq \frac{3}{(\lambda_m(K_{ix}^{-1})\lambda_m(K_{px}))}. \tag{4.57}$$

The derivative of (4.50) is

$$\dot{V} = \dot{x}^T[-C_x\dot{x} - \alpha_f C_{x\theta}\dot{x} - K_{dx}\dot{x} + \sigma_x(0)] + \frac{\mu_x}{2}s_x^T K_{ix}^{-1}s_x + x^T s_x$$

$$+ \frac{\mu_x}{2}\dot{x}^T M_x\dot{x} + \frac{\mu_x}{2}x^T[-C_x\dot{x} - \alpha_f C_{x\theta}\dot{x} - K_{px}x - \sigma_x - s_x + \sigma_x(0)] \tag{4.58}$$

$$- \frac{1}{2}\tilde{w}_r(z_x)\left[\left(\dot{x} + \frac{\mu_x}{2}x\right)^T \phi_r^T(z_x)\right] - \frac{1}{2}\tilde{w}_l(z_x)\left[\left(\dot{x} + \frac{\mu_x}{2}x\right)^T \phi_l^T(z_x)\right]$$

$$+ \frac{1}{2}tr_x\left[\frac{d}{dt}\tilde{w}_r^T(z_x)k_w^{-1}\tilde{w}_r(z_x)\right] + \frac{1}{2}tr_x\left[\frac{d}{dt}\tilde{w}_l^T(z_x)k_w^{-1}\tilde{w}_l(z_x)\right].$$

Using the updating law (4.49),

$$\frac{1}{2}tr_x\left[\frac{d}{dt}\tilde{w}_r^T(z_x)k_w^{-1}\tilde{w}_r(z_x)\right] - \frac{1}{2}\tilde{w}_r(z_x)\left[\left(\dot{x} + \frac{\mu_x}{2}x\right)^T \phi_r^T(z_x)\right] = 0$$

$$\frac{1}{2}tr_x\left[\frac{d}{dt}\tilde{w}_l^T(z_x)k_w^{-1}\tilde{w}_l(z_x)\right] - \frac{1}{2}\tilde{w}_l(z_x)\left[\left(\dot{x} + \frac{\mu_x}{2}x\right)^T \phi_l^T(z_x)\right] = 0.$$

Equation (4.58) becomes

$$\dot{V} = \dot{x}^T[-C_x\dot{x} - \alpha_f C_{x\theta}\dot{x} - K_{dx}\dot{x} + \sigma_x(0)] + \frac{\mu_x}{2}\hat{\xi}_x^T K_{ix}^{-1}\hat{\xi}_x + x^T\hat{\xi}_x$$

$$+ \frac{\mu_x}{2}\dot{x}^T M_x\dot{x} + \frac{\mu_x}{2}x^T[-C_x\dot{x} - \alpha_f C_{x\theta}\dot{x} - K_{px}x - \sigma_x - \hat{\xi}_x + \sigma_x(0)]. \tag{4.59}$$

Now using the property $X^T Y + Y^T X \leq X^T \Lambda X + Y^T \Lambda^{-1} Y$

$$-\frac{\mu_x}{2}x^T C_x\dot{x} \leq \frac{\mu_x}{2}\lambda_M(C_x)\left(x^T x + \dot{x}^T \dot{x}\right)$$

$$-\frac{\mu_x\alpha_f}{2}x^T C_{x\theta}\dot{x} \leq \frac{\mu_x}{2}\lambda_M(\alpha_f C_{x\theta})\left(x^T x + \dot{x}^T \dot{x}\right), \tag{4.60}$$

where $\|C_x\| \leq k_{c_x}$ and $\|C_{x\theta}\| \leq k_{c_{x\theta}}$. So, $s_x = K_{ix}$, $s_x^T K_{ix}^{-1}s_x$ becomes $x^T s_x$, and $x^T s_x$ becomes $x^T K_{ix}$. Now using the Lipschitz condition (4.45),

$$\frac{\mu_x}{2}x^T[\sigma_x(0) - \sigma_x] \leq \frac{\mu_x}{2}k_{\alpha_x}\|x\|^2. \tag{4.61}$$

From (4.60) and (4.61),

$$\dot{V} = -\dot{x}^T \begin{bmatrix} C_x + \alpha_f C_{x\theta} + K_{dx} - \frac{\mu_x}{2} M_x \\ -\frac{\mu_x}{2}\lambda_M(C_x) - \frac{\mu_x}{2}\lambda_M(\alpha_f C_{x\theta}) \end{bmatrix} \dot{x}$$
$$-x^T \left[\frac{\mu_x}{2} K_{px} - K_{ix} - \frac{\mu_x}{2}k_{œx} - \frac{\mu_x}{2}\lambda_M(C_x) - \frac{\mu_x}{2}\lambda_M(\alpha_f C_{x\theta}) \right] x. \tag{4.62}$$

Using (4.44), (4.62) becomes

$$\dot{V} \le -\dot{x}^T \left[\lambda_m(C_x) + \lambda_m(\alpha_f C_{x\theta}) + \lambda_m(K_{dx}) - \frac{\mu_x}{2}\lambda_M(M_x) - \frac{\mu_x}{2}\lambda_M(C_x) - \frac{\mu_x}{2}\lambda_M(\alpha_f C_{x\theta}) \right] \dot{x}$$
$$-x^T \left[\frac{\mu_x}{2}\lambda_m(K_{px}) - \lambda_M(K_{ix}) - \frac{\mu_x}{2}k_{œx} - \frac{\mu_x}{2}\lambda_M(C_x) - \frac{\mu_x}{2}\lambda_M(\alpha_f C_{x\theta}) \right] x. \tag{4.63}$$

So $\dot{V}_x \le 0$, $\|x\|$ minimizes if two conditions are met: (1) $\lambda_m(C_x) + \lambda_m(\alpha_f C_{x\theta}) + \lambda_m(K_{dx}) \ge \frac{\mu_x}{2}[\lambda_M(M_x) + \lambda_M(C_x) + \lambda_M(\alpha_f C_{x\theta})]$; (2) $\lambda_m(K_{px}) \ge \frac{2}{\mu_x}\lambda_M(K_{ix}) + k_{œx} + \lambda_M(C_x) + \lambda_M(\alpha_f C_{x\theta})$. Now using (4.57) and $\lambda_m\left(K_{ix}^{-1}\right) = \frac{1}{\lambda_M(K_{ix})}$, we have

$$\lambda_m(K_{dx}) \ge \frac{1}{4}\sqrt{\frac{1}{3}\lambda_m(M_x)\lambda_m(K_{px})} \left[1 + \frac{\lambda_M(C_x) + \lambda_M(\alpha_f C_{x\theta})}{\lambda_M(M_x)}\right] - \lambda_m(C_x) - \lambda_m(\alpha_f C_{x\theta}) \tag{4.64}$$

again $\frac{2}{\gamma}\lambda_M(K_{ix}) = \frac{2}{3}\lambda_m(K_{px})$. Hence,

$$\lambda_M(K_{ix}) \le \frac{1}{6}\sqrt{\frac{1}{3}\lambda_m(M_x)\lambda_m(K_{px})}\frac{\lambda_m(K_{px})}{\lambda_M(M_x)}. \tag{4.65}$$

Also

$$\lambda_m(K_{px}) \ge \frac{3}{2}[k_{œx} + \lambda_M(C_x) + \lambda_M(\alpha_f C_{x\theta})]. \tag{4.66}$$

Let us assume that there prevails a ball of radius ς in the three-dimensional space. This three-dimensional space is represented by x-, y-, and θ-components. The ball center is at origin of the state-space system, where $\dot{V}_x \le 0$, $\dot{V}_y \le 0$, $\dot{V}_\theta \le 0$. The origin of the closed-loop systems represented by (4.43) is stable equilibrium. Similarly, the closed-loop system along other components will be in stable equilibrium. Now we will prove for the asymptotic stability of the origin. For that, we use La Salle's theorem by defining the term Π_x, Π_y and Π_θ as follows:

$$\Pi_x = \left\{ z_x(t) = \left[x^T, \dot{x}^T, \xi_x^T\right]^T \in \Re^{3n} : \dot{V}_x = 0 \right\}, \xi_x \in \Re^n, x = 0 \in \Re^n, \dot{x} = 0 \in \Re^n$$
$$\Pi_y = \left\{ z_y(t) = \left[y^T, \dot{y}^T, \xi_y^T\right]^T \in \Re^{3n} : \dot{V}_y = 0 \right\}, \xi_y \in \Re^n, y = 0 \in \Re^n, \dot{y} = 0 \in \Re^n$$
$$\Pi_\theta = \left\{ z_\theta(t) = \left[\theta^T, \dot{\theta}^T, \xi_\theta^T\right]^T \in \Re^{3n} : \dot{V}_\theta = 0 \right\}, \xi_\theta \in \Re^n, \theta = 0 \in \Re^n, \dot{\theta} = 0 \in \Re^n. \tag{4.67}$$

Using (4.58) and substituting $x = 0$ and $\dot{x} = 0$, we have $\dot{V}_x = 0$. Similar analysis with $y = 0$ and $\dot{y} = 0$ and also $\theta = 0$ and $\dot{\theta} = 0$ will yield $\dot{V}_y = 0$ and $\dot{V}_\theta = 0$.

Similarly, these conditions hold good for $\dot{x} = 0$, $\dot{y} = 0$ and $\dot{\theta} = 0$ for all $t \geq 0$. Therefore $z_x(t)$, $z_y(t)$, and $z_\theta(t)$ belong to Π_x, Π_y, and Π_θ, respectively. Also, imparting these conditions to (4.43), we have $\dot{\xi}_x = 0$, $\dot{\xi}_y = 0$ and $\dot{\xi}_\theta = 0$. Also, $\xi_x = 0$, $\xi_y = 0$ and $\xi_\theta = 0$ for all $t \geq 0$. So $z_x(t)$ is the only initial condition in Π_x, $z_y(t)$ is the only initial condition in Π_y, and $z_\theta(t)$ is the only initial condition in Π_θ. Therefore, origin is asymptotically stable according to La Salle's theorem. Now for global stability of the closed-loop system mentioned by (4.43), the following conditions needs to be met: $\lim_{t \to \infty}^x(t) = 0$, when the initial condition of $[x, \dot{x}, \xi_x]$ is inside of Π_x. $\lim_{t \to \infty}^y(t) = 0$, when the initial condition of $[y, \dot{y}, \xi_y]$ is inside of Π_y. $\lim_{t \to \infty}^\theta(t) = 0$, when the initial condition of $[\theta, \dot{\theta}, \xi_\theta]$ is inside of Π_θ

4.4 Experimental Results

In order to analyze and validate the bidirectional type-2 fuzzy PD and PID controllers, a two-floor building structure is designed and constructed. The detailed structure and the placements of actuators are mentioned in Chap. 3. The relative acceleration in the second floor is subtracted by the ground floor acceleration. Numerical integrators are used to compute the velocity and position from the accelerometer signal. Since there is not angular sensor, the angular accelerations are calculated by (3.70) and (3.71) mentioned in Chap. 3, where $\ddot{\theta}_1$ and $\ddot{\theta}_2$ are the angular accelerations of the first and second floors. The theorems of this chapter give the sufficient conditions of minimal proportional and derivative gains and maximum integral gain. To compare with the other algorithm in the same condition, all PID gains are the same. The upper bounds and lower bounds of the structure model are

$$\lambda_M(M_x) = 10, \lambda_M(M_y) = 10$$
$$\lambda_m(k_{c_x}) = 20, \lambda_m(k_{c_{x\theta}}) = 8\lambda_m(k_{c_y}) = 22, \lambda_m(k_{c_{y\theta}}) = 6 \qquad (4.68)$$
$$\lambda_{j_t}(J_0) = 5, \lambda_m(k_{c_\theta}) = 21, \lambda_m(k_{c_{x\theta}}) = 8, \lambda_m(k_{c_{y\theta}}) = 6.$$

k_{α_x} is affected by the external force $\mathbf{F}$. The maximum force to actuate the building structure prototype during experiment is 300 N. Therefore, we select $k_{\alpha_x} = 400$. Y- and θ-components are extracted with the similar method. For X-component, ranges of the PID gains are

$$\lambda_m(K_{px}) \geq 1247, \lambda_m(K_{dx}) \geq 50, \lambda_M(K_{ix}) \leq 2677, \lambda_m(K_{py}) \geq 1247$$
$$\lambda_m(K_{dy}) \geq 50, \lambda_M(K_{iy}) \leq 2677, \lambda_{j_0}(K_{p\theta}) \geq 1431, \lambda_{j_0}(K_{d}) \geq 75, \lambda_{J_0}(K_{i\theta}) \leq 3789.$$
$$(4.69)$$

We first use type-2 fuzzy logic system toolbox [21] to design the type-2 fuzzy system. Due to its iterative nature, the computational cost of the calculation of the type-2 fuzzy system output is big [22]. In order to tackle with these situations, several TR methods have been proposed for reducing the computational cost of the type-2 fuzzy inference mechanism. The KM algorithms are iterative procedures widely used in fuzzy logic theory. They are known to converge monotonically and super exponentially fast; however, several (usually two to six) iterations are still needed before convergence occurs [17]. Wu categorized the TR methods as enhancements to the KMs, which improved the computational cost of the KM, and alternative TR methods, which are closed-form approximations to the KM algorithm [23]. KM method is most popular due its novelty and adaptiveness [24]. The type-reduction and the defuzzification methods supported by type-2 fuzzy logic system toolbox are (1) KM algorithm, (2) enhanced KM algorithm (EKM), (3) iterative algorithm with stop condition (IASC), (4) enhanced IASC (EIASC), (5) enhanced opposite direction searching algorithm (EODS), (6) Wu–Mendel uncertainty bound method (WM), (7) Nie–Tan method (NT), And (8) Begian–Melek–Mendel method (BMM). In type-2 fuzzy logic system toolbox, it is possible to state the antecedent MFs with the MF types that already prevail in the MATLAB Fuzzy Logic Toolbox. Hence, it is feasible to implement the MATLAB functions of LMF and UMF in the same pattern. But there is an additional parameter associated to each type of MFs that illustrates the height of the corresponding MF. For example, a triangle MF is stated having the parameters $l_{t2}, c_{t2}, r_{t2}, h_{t2}$ which defines the left point, the center point, right point, and the height of the MF, respectively. The parameter h_{t2} is generally utilized to develop FOU in the type-2 fuzzy systems, most specifically in type-2 fuzzy controller design. The control performance is evaluated to minimize the relative displacement of each floor of the building. The membership functions are Gaussian functions, and they are designed by the method of [3], see Fig. 4.1, where mfU and mfL are the upper and lower membership functions. The advantages of Gaussian membership functions are as follows: it is simpler in design because they are easier to represent and optimize, always continuous, and faster for small rule bases. Gaussian type-2 fuzzy logic systems are faster than the corresponding trapezoidal type-2 fuzzy logic systems when the same number of MFs and the same type-reduction and defuzzification method are used. Since small rule bases are usually used in practice, Gaussian type-2 fuzzy logic systems seem more favorable in terms of computational cost [25]. For the floor position and velocity, we use three linguistic variables and three membership functions; they are normalized in $[-1, 1]$. KM method [10] is utilized to defuzzify the type-2 fuzzy system. For the formulation of type-2 fuzzy rules, the FIS variables are selected as input variables (position error and velocity error) and output variable (control force). IF-THEN rules are applied. IF and AND conditions are applied between position error and velocity error, whereas THEN condition gives the required control force. Considering X-component, 15 fuzzy rules are applied. Also for Y- and θ-components, similar set of 15 fuzzy rules are applied, respectively. We find 15 fuzzy rules are sufficient to maintain minimum regulation errors. For design purpose, we choose $\mu_x = 6$. We also find that for the type-1 fuzzy, at least nine fuzzy rules are needed to have the similar regulation errors as the type-2 fuzzy system.

The signal to the shake table is the Northridge earthquake. The displacement is scaled from 16.92 to 1.50 cm, and the time is scaled from 40 to 12 s. The control

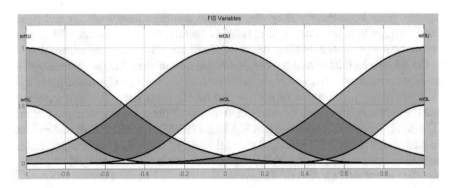

Fig. 4.1 Upper and lower bounds of the membership functions

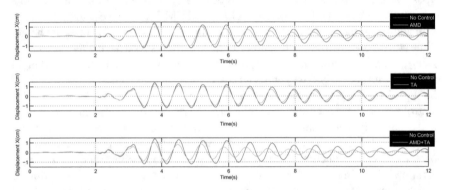

Fig. 4.2 PD control in X-direction

object is to minimize the relative displacement of each floor in bidirection. From (4.69), the PID control gains are

$$K_{px} = 1800, \; K_{py} = 2000, \; K_{p\theta} = 2200,$$
$$K_{dx} = 160, \; K_{dy} = 220, \; K_{d\theta} = 300, \; K_{ix} = 2000, \; K_{iy} = 2300, \; K_{i\theta} = 3500.$$

$$(4.70)$$

The PD control gains are $K_{px} = 1800, K_{py} = 2000, K_{p\theta} = 2200, K_{dx} = 160 K_{dy} = 220, K_{d\theta} = 300$.

We compare our control with classical PD/PID, type-1 fuzzy PD/PID in three cases: (1) without any active control (No Control); (2) with the TA; and (3) with both the AMD and the TA (AMD+TA). The results of these controllers are shown in Figs. 4.2, 4.3, 4.4, 4.5, 4.6, and 4.7. The control signals of type-2 fuzzy PD and PID are displayed in Figs. 4.8 and 4.9. We define the average vibration displacement as $MSE = \frac{1}{N} \sum_{k=1}^{N} x(k)^2$, $x(k)$ is the displacement of the floor, and N is the total data number. The comparison results of the average vibration displacement are shown in Tables 4.1, 4.2, 4.3, 4.4, 4.5, 4.6, 4.7, 4.8, and 4.9. Here, ↓ sign indicates decrease.

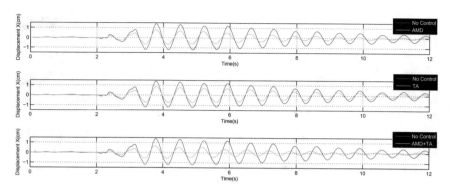

Fig. 4.3 PID control in X-direction

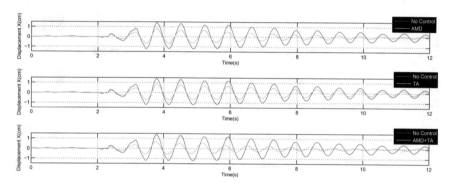

Fig. 4.4 Type-1 fuzzy PD control in X-direction

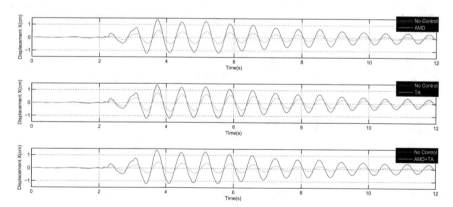

Fig. 4.5 Type-1 fuzzy PID control in X-direction

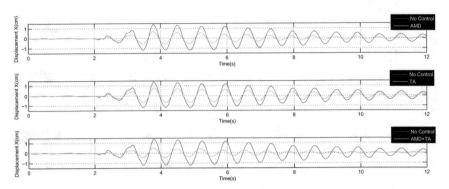

Fig. 4.6 Type-2 fuzzy PD control in X-direction

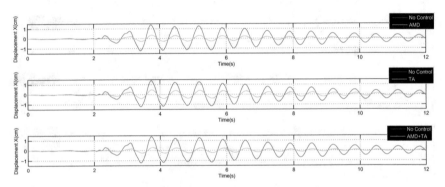

Fig. 4.7 Type-2 fuzzy PID control in X-direction

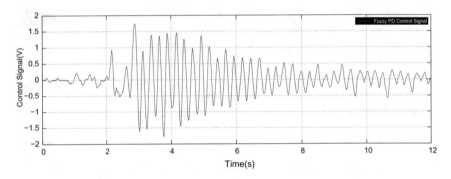

Fig. 4.8 Control signal of type-2 fuzzy PD

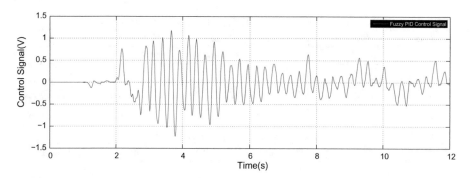

Fig. 4.9   Control signal of type-2 fuzzy PID

Table 4.1   Average vibration displacement by AMD+TA

Direction	PD control	% ↓ error	PID control	% ↓ error	No control
X	0.2987	60.2	0.2373	68.5	0.7514
Y	0.0719	45.53	0.0783	59.3	0.1320
θ	0.0696	40.8	0.0611	47.6	0.1174

Table 4.2   Average vibration displacement with PD control

Direction	With AMD	% ↓ error	With TA	% ↓ error	No control
X	0.4832	35.7	0.5802	22.78	0.7514
Y	0.0981	25.68	0.1012	23.3	0.1320
θ	0.0902	23.1	0.0801	31.7	0.1174

Table 4.3   Average vibration displacement with PID control

Direction	With AMD	% ↓ error	With TA	% ↓ error	No control
X	0.3632	51.6	0.4911	34.6	0.7514
Y	0.0849	35.6	0.0969	26.5	0.1320
θ	0.0811	30.8	0.0713	40.0	0.1174

Table 4.4   Average vibration displacement by AMD+TA

Direction	Type-1 fuzzy PD	% ↓ error	Type-1 fuzzy PID	% ↓ error	No control
X	0.1843	75.4	0.1590	78.8	0.7514
Y	0.0531	60.0	0.0488	63.3	0.1320
θ	0.0578	50.7	0.0423	63.9	0.1174

Table 4.5 Average vibration displacement with Type-1 Fuzzy PD control

Direction	With AMD	% ↓ error	With TA	% ↓ error	No control
X	0.2121	71.7	0.3398	54.7	0.7514
Y	0.0633	52.4	0.0733	44.5	0.1320
θ	0.0820	30.1	0.0674	42.6	0.1174

Table 4.6 Average vibration displacement with Type-1 Fuzzy PID control

Direction	With AMD	% ↓ error	With TA	% ↓ error	No control
X	0.1944	74.1	0.2416	67.8	0.7514
Y	0.0581	55.9	0.0634	51.96	0.1320
θ	0.0711	40.0	0.0481	59.1	0.1174

Table 4.7 Average vibration displacement by AMD+TA

Direction	Type-2 fuzzy PD	% ↓ error	Type-2 fuzzy PID	% ↓ error	No control
X	0.1348	82.1	0.1121	85.1	0.7514
Y	0.0320	75.7	0.0299	77.3	0.1320
θ	0.0395	66.35	0.0278	76.3	0.1174

Table 4.8 Average vibration displacement with type-2 fuzzy PD control

Direction	With AMD	% ↓ error	With TA	% ↓ error	No control
X	0.1693	77.4	0.2339	68.8	0.7514
Y	0.0433	67.1	0.0670	49.2	0.1320
θ	0.0489	58.3	0.0396	66.2	0.1174

Table 4.9 Average vibration displacement with type-2 fuzzy PID control

Direction	With AMD	% ↓ error	With TA	% ↓ error	No control
X	0.1575	79.1	0.2080	72.3	0.7514
Y	0.0389	70.5	0.0584	55.7	0.1320
θ	0.0431	63.2	0.0371	68.3	0.1174

We can see that all PD/PID, type-1 fuzzy PD/PID, and type-2 fuzzy PD/PID controllers work well with AMD and TA, because both AMD and TA act simultaneously. The vibration attenuations along X- and Y-directions are much better than θ-direction, because the active vibration control is achieved by the position of the actuator and the torque direction of the actuator. PID controller is better than PD to minimize the vibration in all three directions. The active control of structures can be improved by adding the fuzzy compensation. The type-2 fuzzy controller is capable of providing more vibration attenuation than the type-1 fuzzy controller. The best results are the type-2 fuzzy PID control.

4.5 Conclusion

In this chapter, type-2 fuzzy PD and PID controls for building structures with AMD and TA are proposed. By utilizing the Lyapunov theory, sufficient conditions of stability are extracted to tune PD/PID gains. The mentioned methodology is successfully implemented in a two-storey building prototype. The experimental results show that the type-2 fuzzy PD/PID controllers work better than type-1 fuzzy PD/PID for the AMD and TA.

References

1. K.M. Choi, S.W. Cho, D.O. Kim, I.W. Lee, Active control for seismic response reduction using modal-fuzzy approach. Int. J. Solids Struct. **42**(16–17), 4779–4794 (2005)
2. D.G. Reigles, M.D. Symans, Supervisory fuzzy control of a base-isolated benchmark building utilizing a neuro-fuzzy model of controllable fluid viscous dampers, in *Structural Control and Health Monitoring,* vol. 13, no. 2–3 (2006), pp. 724–747
3. R. Guclu, Sliding mode and PID control of a structural system against earthquake. Math. Comput. Model. **44**(1–2), 210–217 (2006)
4. K.S. Park, H.M. Koh, S.Y. Ok, Active control of earthquake excited structures using fuzzy supervisory technique. Adv. Eng. Softw. **33**(11–12), 761–768 (2002)
5. D. Das, T.K. Datta, A. Madan, Semiactive fuzzy control of the seismic response of building frames with MR dampers. Earthq. Eng. Struct. Dyn. **41**, 99–118 (2012)
6. N.D. Duc, N.l. Vu, D.T. Tran, H.L. Bui, A study on the application of hedge algebras to active fuzzy control of a seism-excited structure. J. Vibr. Control **18**(14), 2180–2200 (2012)
7. S.F. Ali, A. Ramaswamy, Optimal fuzzy logic control for MDOF structural systems using evolutionary algorithms. Eng. Appl. Artif. Intell. **22**, 407–419 (2009)
8. A.S. Ahlawat, A. Ramaswamy, Multiobjective optimal FLC driven hybrid mass damper system for torsionally coupled, seismically excited structures. Earthq. Eng. Struct. Dyna. **31**(12), 2121–2139 (2002)
9. H.-L. Bui, T.A. Le, V.B. Bui, Explicit formula of hedge-algebras-based fuzzy controller and applications in structural vibration control. Appl. Soft Comput. **60**, 150–166 (2017)
10. Q. Liang, J.M. Mendel, Interval type-2 fuzzy logic systems: theory and design. IEEE Trans. Fuzzy Syst. **8**(5), 535–550 (2002)
11. R. Sepúlveda, O. Castillo, P. Melin, A. Rodríguez-Díaz, O. Montiel, Experimental study of intelligent controllers under uncertainty using type-1 and type-2 fuzzy logic. Inf. Sci. **177**(10), 2023–2048 (2007)
12. T.C. Lin., H.L. Liu, M.J. Kuo, Direct adaptive interval type-2 fuzzy control of multivariable nonlinear systems. Eng. Appl. Artif. Intell. **22**(3), 420–430 (2009)
13. H. Shariatmadar, S. Golnargesi, M.R. Akbarzadeh-T, Vibration control of buildings using ATMD against earthquake excitations through interval type-2 fuzzy logic controller. Asian J. Civil Eng. **15**(3), 328–331 (2014)
14. A. Bathaei, S.M. Zahrai, M. Ramezani, Semi-active seismic control of an 11-DOF building model with TMD+MR damper using type-1 and -2 fuzzy algorithms. J. Vibr. Control (2017)

15. R.I. John, S. Coupland, type-2 fuzzy logic: a historical view. IEEE Comput. Intell. Mag. **2**(1), 57–62 (2007)
16. J.M. Mendel, *Uncertain Rule-based Fuzzy Logic Systems: Introduction and New Directions* (Prentice Hall PTR, Upper Saddle River, NJ, 2001)
17. D. Wu, J.M. Mendel: Enhanced Karnik-Mendel algorithms. IEEE Trans. Fuzzy Syst. **17**(4), 923–934 (2009)
18. B. Bernhard, C.J. Mulvey, A constructive proof of the Stone-Weierstrass theorem. J. Pure App. Algebra **116**(1–3), 25–40 (1997)
19. E.D. Sontag, Y. Wang, On characterizations of the input-to-state stability property. Syst. Control Let. **24**(5), 351–359 (1995)
20. F.L. Lewis, D.M. Dawson, C.T. Abdallah, *Robot Manipulator Control: Theory and Practice,* 2nd edn. (Marcel Dekker Inc., 2004)
21. A. Taskin, T. Kumbasar, An open source MATLAB/Simulink toolbox for interval type-2 fuzzy logic systems, in *IEEE Symposium Series on Computational Intelligence* (2015)
22. D. Wu, M. Nie, Comparison and practical implementation of type reduction algorithms for type-2 fuzzy sets and systems, in *IEEE International Conference Fuzzy Systems,* Taipei, Taiwan (2011)
23. D. Wu, Approaches for reducing the computational cost of interval type-2 fuzzy logic systems: overview and comparisons. IEEE Trans. Fuzzy Syst. **21**(1), 80–99 (2013)
24. D. Wu, On the fundamental differences between type-1 and interval type-2 fuzzy logic controllers. IEEE Trans Fuzzy Syst. **10**(5), 832–848 (2012)
25. D. Wu, Twelve considerations in choosing between Gaussian and trapezoidal membership functions in interval type-2 fuzzy logic controllers, in *IEEE International Conference on Fuzzy Systems,* Brisbane, QLD (2012), pp. 1–8

Chapter 5
Discrete-Time Fuzzy Sliding-Mode Control

5.1 Introduction

Active vibration control of building structures under earthquake loadings is a popular field among civil and mechanical engineers. Different control devices and algorithms were proposed and implemented in the last few decades [1, 2]. One of the main challenges in the structural control design is the presence of uncertainties in the building structures, especially in parametric level. Robust control is a well-established technique, which can deal with these uncertainties and disturbances present in the real systems like the building structures.

Research reveals that sliding-mode control (SMC) is considered to be an effective robust control strategy for uncertain systems. The SMC is designed for uncertain nonlinear systems [3]. It is very much effective in terms of robustness against the changes in the parameters and external disturbances. It has been successfully applied for structural control [4]. In [5], SMC is used to control bending and torsional vibration of a six-storey flexible structure. A new robust control system for an active tuned mass damper is implemented in a high-rise building. The controller is a blended innovation of two-loop sliding-mode controller with a dynamic state predictor [6]. In [7], an active vibration control for a two-storeyed flexible structure was proposed where the SMC is designed utilizing linear quadratic regulator (LQR) approach in order to validate stable motion while undergoing sliding. An approach related to adaptive fuzzy sliding-mode in order to eliminate the damage of the nonlinear structure was suggested by [8].

The implementation of digital computers and samplers in the field of control systems has popularized the research of discrete-time systems. A necessary and sufficient condition for discrete-time sliding motion is suggested in [9] as follows:

$$| S(k+1) | < | S(k) | .$$

© The Author(s), under exclusive license to Springer Nature Switzerland AG 2020
W. Yu and S. Paul, *Active Control of Bidirectional Structural Vibration*,
SpringerBriefs in Applied Sciences and Technology,
https://doi.org/10.1007/978-3-030-46650-3_5

One more important condition for discrete-time sliding-mode with the consideration that sliding hyperplane $S(k) = 0$ should satisfy [10]

$$\mid S(k) \mid < g,$$

where the parameter $g > 0$ is termed as quasi-sliding-mode band (QSMB) width. However, owing to a finite sampling frequency characteristic in discrete-time systems, the system states can only be expected to approach the selected sliding surface and remain around it, instead of remaining on the surface when the system undergoes external disturbances. Therefore, the so-called quasi-sliding-mode (QSM) concept was introduced and discussed in discrete-time systems [10, 11]. The system states are required to monotonically approach the sliding surface until they enter the vicinity of the surface, and they then remain inside. The vicinity of the sliding surface is called QSMB. Under this QSM definition, it is noted that the system states are not required to cross the sliding surface, as in the definition given in [12]. The undesirable chattering and high-frequency switching between different values of the control signal are avoided. Since its state does not have to cross the sliding hyperplane in each control step, the control strategy can be linear and, consequently, the undesirable chattering is avoided. The strategies guarantee improved robustness, faster transient response, and better steady-state accuracy of the controlled system [10].

In general cases, discrete-time control or sampling control is most suited for the structural control. The sampling period is considered to be the important feature that plays a significant role in the performance of the control system. In [13], a discrete-time variable structure control strategy on the basis of discrete reaching law method in order to minimizing the dynamic responses of seismically excited structures was suggested. A time-delayed discrete-time variable structure control method in order to mitigate vibration in the linear structures was proposed by [14]. A novel discrete-time variable structure control method in combination with fuzzy adaptive regulation for seismically excited linear structure with the intention of subsiding heavy chattering effect was presented by [15]. An adaptive backstepping fuzzy sliding-mode control for the approximation of unknown system dynamics and vibration control associated with cantilever beam is proposed by [16]. In [17], a new discrete-time variable structure control method incorporated with discrete-time composite reaching law was proposed for vibration attenuation in seismically excited linear structure. An innovative methodology of adaptive hyperbolic tangent sliding-mode control for the control of structural vibration under uncertain earthquake loads is proposed in [18]. In this chapter, we present a fuzzy discrete sliding-mode control (FDSMC) for the minimization of structural vibration along all three components under the effect of bidirectional earthquake forces. The analysis is based on the lateral–torsional vibration under the bidirectional waves. Also in order to reduce chattering, the suggested FDSMC with time-varying gain is effective. We prove that the closed-loop system with SMC and fuzzy identifier is uniformly stable by utilizing the Lyapunov stability theorem. The experimental results and analysis using the FDSMC validate its effectiveness and stability. Finally, the results are compared with standard discrete sliding-mode controller (DSMC) and PID controller to verify the superior performance of FDSMC in mitigating the earthquake vibrations.

5.2 Discrete-Time Model of Building Structure

The continuous-time model of n-floor building structure under bidirectional external forces is

$$M\ddot{\mathbf{x}}(t) + C\dot{\mathbf{x}}(t) + F_s(\mathbf{x}) + f_e(t) = \Gamma u(t), \tag{5.1}$$

where $\mathbf{x} \in \Re^{3n}$, $\mathbf{x} = [x_1 \cdots x_n, y_1 \cdots y_n, \theta_1 \cdots \theta_n]^T$, $\mathbf{x}$ is the displacement, $M \in \Re^{3n \times 3n}$ is the mass, $C \in \Re^{3n \times 3n}$ the damping coefficient, $F_s = [f_{s,1} \cdots f_{s,n}] \in \Re^{3n}$ is the structure stiffness force vector, and $f_e \in \Re^{3n}$ is the bidirectional external force applied to the structure, $f_e = [f_x, f_y, 0 \cdots 0]^T$, $u \in \Re^{3n}$ is the control signals which is fed to the dampers. The structure stiffness can be model as

$$F_s(\mathbf{x}) = f_s(\mathbf{x}) + \Gamma d_u, \tag{5.2}$$

where $f_s(\mathbf{x})$ is the structure stiffness force; it can be modeled as a linear model $f_s = K\mathbf{x}$, or a nonlinear model, and d_u is the damping and friction force vector of the dampers. We define $z_1(t) = \mathbf{x}$ and $z_2(t) = \dot{\mathbf{x}}$; the model (5.1) can be transformed into the following state-space model:

$$\dot{Z}(t) = Az(t) + Bu(t) + F_s(z) + f_e(t), \tag{5.3}$$

where $z(t) = \begin{bmatrix} z_1(t) \\ z_2(t) \end{bmatrix}$, $A = \begin{bmatrix} 0 & 0 \\ 0 & -M^{-1}C \end{bmatrix}$, $B = \begin{bmatrix} 0 \\ M^{-1}\Gamma \end{bmatrix}$, $F_s(z) = M^{-1}F_s(z)$, $f_e(t) = M^{-1}f_e(t)$.

$F_s(\mathbf{x})$ and $f_e(t)$ can be regarded as the uncertainty parts of the linear system $\dot{Z} = Az + Bu$. Clearly, in the absence of the external forces, the building structure is stable. So it is reasonable to assume that $F_s(z)$ is bounded, $\|F_s(z)\| \leqslant d_s$. The external forces are bounded, $\|f_e(t)\| \leqslant d_e$. In order to discretize the continuous-time model, we assume that the control force and the external forces are constant during the sampling period T, i.e.,

$$u(t) = u(kT), \quad f_e(t) = f_e(kT), \quad kT \leqslant t \leqslant (k+1)T.$$

The discrete-time model of (5.1) is [13],

$$z(k+1) = A_d + B_d u(k) + F_{ds}[z(k)] + f_{de}(k), \tag{5.4}$$

where $z(k)$ is a state vector, A_d is a state matrix, $A_d = e^{AT}$, B_d is the input vector, $B_d = \left[\int e^{A\tau} d\tau\right] B$, $u(k)$ is a scalar input, $F_{ds}(k)$ is the model uncertainty matrix, and $f_{de}(k)$ is the excitation. Since A_d and B_d are unknown, (5.4) is written as the following general nonlinear model:

$$z(k+1) = f[z(k)] + g[z(k)] u(k) + d[z(k)], \tag{5.5}$$

where $f[z(k)] = A_d z(k)$, $g[z(k)] = \Gamma_{i,j} B_d$, $d[z(k)] = F_{ds}[z(k)] + f_{de}(k)$, Γ is defined as the location matrix of the dampers,

$$\Gamma_{i,j} = \begin{cases} 1 & \text{if } i = j = s \\ 0 & \text{otherwise} \end{cases}, \tag{5.6}$$

where $\forall i, j \in \{1, \dots, n\}, s \subseteq \{1, \dots, n\}$, s are the floors on which the dampers are installed. For a two-floor building, $\Gamma = \begin{bmatrix} \Gamma_{1,1} & \Gamma_{1,2} \\ \Gamma_{2,1} & \Gamma_{2,2} \end{bmatrix}$. If the damper is placed on the second floor, $\Gamma = \begin{bmatrix} 0 & 0 \\ 0 & 1 \end{bmatrix}$.

5.3   Fuzzy Modeling of Structure

We use the following fuzzy system to modeling the unknown nonlinear functions $f[z(k)]$, $g[z(k)]$, and $d[z(k)]$ in (5.5). The unknown nonlinear functions f and g are approximated as

$$\begin{aligned} f[z(k)] + d[z(k)] &= \hat{f} + \varepsilon_f \\ g[z(k)] &= \hat{g} + \varepsilon_g, \end{aligned} \tag{5.7}$$

where ε_f and ε_g are the modeling errors, and $\hat{f}$ and $\hat{g}$ are the estimations of $f[z(k)] + d[z(k)]$ and $g[z(k)]$. We use the following fuzzy system to model $f[z(k)]$, $g[z(k)]$, and $d[z(k)]$. For pth fuzzy rules for $f[z(k)]$, $g[z(k)]$, and $d[z(k)]$,

R^i: IF $(x_i$ is $A_{1i})$ and $(y_i$ is $A_{2i})$ and $(\theta_i$ is $A_{3i})$ and $(\dot{x}_i$ is $A_{4i})$ and $(\dot{y}_i$ is $A_{5i})$ and $(\dot{\theta}_i$ is $A_{6i})$ THEN $f[z(k)] + d[z(k)]$ is B_{1i}

$$\tag{5.8}$$

R^i: IF $(x_i$ is $A_{1i})$ and $(y_i$ is $A_{2i})$ and $(\theta_i$ is $A_{3i})$ and $(\dot{x}_i$ is $A_{4i})$ and $(\dot{y}_i$ is $A_{5i})$ and $(\dot{\theta}_i$ is $A_{6i})$ THEN $g[z(k)]$ is B_{2i},

$$\tag{5.9}$$

where $A_{1i} \dots$, A_{6i}, B_{1i}, B_{3i}, and B_{3i} are the fuzzy sets. Now by product inference, center-average defuzzification, and a singleton fuzzifier, the output of the fuzzy logic system can be expressed as [19]

$$\hat{F}p = \frac{\left(\sum_{i=1}^{l} w_{pi}[\Pi_{j=1}^{n}\mu_{A_{ji}}]\right)}{\left(\sum_{i=1}^{l}[\Pi_{j=1}^{n}\mu_{A_{ji}}]\right)} = \sum_{i=1}^{l} w_{pi}\sigma_i, \tag{5.10}$$

where $\mu_{A_{ji}}$ is the membership functions of the fuzzy sets A_{ji}, w_{pi} is the point at which $\mu_{\beta_{ji}} = 1$, if we define

$$\sigma_i = \frac{\Pi_{j=1}^{n}\mu_{A_{ji}}}{\sum_{i=1}^{l}\Pi_{j=1}^{n}\mu_{A_{ji}}}. \tag{5.11}$$

The Gaussian functions are chosen as the membership functions as follows:

$$\mu_{A_{ji}} = \exp\left(-\frac{(x_j - c_{ji})^2}{\rho_{ji}^2}\right), \tag{5.12}$$

where c_{ji} and ρ_{ji} are the mean and variance of the Gaussian function, respectively. In the matrix form, (5.10) can be expressed as

$$\hat{F}p = w(k)\sigma[z(k)], \tag{5.13}$$

where

$$w(k) = \begin{bmatrix} w_{11}(k) & & w_{1l}(k) \\ & \ddots & \\ & & \ddots \\ w_{m1}(k) & & w_{ml}(k) \end{bmatrix} \varepsilon R^{m\times l}$$

also $\sigma[z(k)] = [\sigma_1 \cdots \sigma_l]\varepsilon R^{l\times 1}$. Now using (5.13) and since $\hat{f}$ and $\hat{g}$ are the estimations of $f[z(k)] + d[z(k)]$ and $g[z(k)]$ then

$$\begin{aligned} \hat{f} &= w_f(k)\sigma_f[z(k)] \\ \hat{g} &= w_g(k)\sigma_g[z(k)]. \end{aligned} \tag{5.14}$$

According to the Stone–Weierstrass Theorem [20], the unknown nonlinear functions f and g are approximated as

$$\begin{aligned} f &= w_f^*(k)\sigma_f[z(k)] + \varepsilon_f \\ g &= w_g^*(k)\sigma_g[z(k)] + \varepsilon_g. \end{aligned} \tag{5.15}$$

The nonlinear system (5.5) can be modeled with fuzzy system as

$$\beta\hat{z}(k+1) = \hat{f}[z(k)] + \hat{g}[z(k)]u(k), \tag{5.16}$$

where β is a positive constant and $\beta > 1$ which is a design parameter. We define the modeling error as

$$e_i(k+1) = \hat{z}(k+1) - z(k+1) \tag{5.17}$$

and

$$\tilde{f} = \hat{f} - f[z(k)] - d[z(k)]$$
$$\tilde{g} = \hat{g} - g[z(k)]. \tag{5.18}$$

Now from (5.5), the nonlinear model can be represented in the fuzzy form as follows:

$$\beta z(k+1) = \left[w_f^*(k)\sigma_f[z(k)] + \varepsilon_f\right] + \left[w_g^*(k)\sigma_g[z(k)] + \varepsilon_g\right]u(k)$$
$$\beta z(k+1) = w_f^*(k)\sigma_f[z(k)] + w_g^*(k)\sigma_g[z(k)]u(k) + \varepsilon_f + \varepsilon_g u(k). \tag{5.19}$$

Now from the Taylor series formula, we have for n variables

$$f(x_1, \ldots, x_n) = \sum_{j=0}^{\infty} \left[\frac{1}{j!}\left(\sum_{k=1}^{n}(x_k - a_k)\frac{\partial}{\partial \acute{x}_k}\right)^j f(\acute{x}_1, \ldots, \acute{x}_n)\right]\acute{x}_1 \tag{5.20}$$

$$= x_1 \ldots \acute{x}_n = x_n.$$

Now applying the Taylor series to the smooth functions $\hat{f}$ and $\hat{g}$, we have

$$\hat{f} = w_f^*(k)\sigma_f[z(k)] + [w_f(k) - w_f^*(k)]\frac{\partial \hat{f}}{\partial[w_f(k)]} + R_f \tag{5.21}$$
$$\hat{f} = w_f(k)\sigma_f[z(k)] + R_f$$

$$\hat{g} = w_g^*(k)\sigma_g[z(k)] + [w_g(k) - w_g^*(k)]\frac{\partial \hat{g}}{\partial[w_g(k)]} + R_g$$
$$\hat{g} = w_g(k)\sigma_g[z(k)] + R_g, \tag{5.22}$$

where $\frac{\partial \hat{f}}{\partial[w_f(k)]} = \sigma_f[z(k)]$ and $\frac{\partial \hat{g}}{\partial[w_g(k)]} = \sigma_g[z(k)]$. Also, R_f and R_g are the remainders of the Taylor formula. Now using (5.18), we can demonstrate

$$\tilde{f} = (w_f(k) - w_f^*(k))\sigma_f[z(k)] + (R_f - \varepsilon_f) \tag{5.23}$$
$$\tilde{f} = \tilde{w}_f(k)\sigma_f[z(k)] + \xi_f$$

$$\tilde{g} = (w_g(k) - w_g^*(k))\sigma_g[z(k)] + (R_g - \varepsilon_g) \tag{5.24}$$
$$\tilde{g} = \tilde{w}_g(k)\sigma_g[z(k)] + \xi_g,$$

where $\tilde{w}_f(k) = w_f(k) - w_f^*(k)$, $\tilde{w}_g(k) = w_g(k) - w_g^*(k)$, $\xi_f = R_f - \varepsilon_f$ and $\xi_g = R_g - \varepsilon_g$. The error dynamics can be expressed using (5.16) and (5.19) as

$$\beta e_i(k+1) = \tilde{w}_f(k)\sigma_f[z(k)] + \tilde{w}_g(k)\sigma_g[z(k)]u(k) + \xi_f + \xi_g u(k). \quad (5.25)$$

In the sake of assuring stability of identification and non-singularity in the controller, the following updating laws are implemented:

$$\Delta w_f(k) = -\eta(k)\sigma_f[z(k)]e_i^T(k)$$
$$\Delta w_g(k) = -\eta(k)u(k)\sigma_g[z(k)]e_i^T(k). \quad (5.26)$$

Theorem 5.1 *If we use fuzzy model (5.16) to identify nonlinear system (5.5) having the updating law given by (5.26), then the identification error $e_i(k)$ is bounded and it satisfies the following relation:*

$$\lim_{k\to\infty} \| e_i(k) \|^2 = \frac{\| \bar{\xi}(k) \|^2 [1 + \pi(k)]}{1 + (1 - \eta(k))\pi(k)} \quad (5.27)$$

provided the dead zone guarantees $\beta \| e_i(k+1) \| > \| e_i(k) \|$ and $0 < \eta(k) < 1, \pi(k) \geq 0$.

Proof We select the following Lyapunov candidate function $V(k)$:

$$V(k) = tr[\tilde{w}_f^T(k)\tilde{w}_f(k)] + tr[\tilde{w}_g^T(k)\tilde{w}_g(k]$$
$$= \sum_{i=1}^n \tilde{w}_f(k)^2 + \sum_{i=1}^n \tilde{w}_g(k)^2 \quad (5.28)$$
$$= \| \tilde{w}_f(k) \|^2 + \| \tilde{w}_g(k) \|^2 .$$

Now we know $\Delta V(k) = V(k+1) - V(k)$. Using this and (5.28)

$$\Delta V(k) = [\| \tilde{w}_f(k+1) \|^2 - \| \tilde{w}_f(k) \|^2] + [\| \tilde{w}_g(k+1) \|^2 - \| \tilde{w}_g(k) \|^2]. \quad (5.29)$$

Now from the updating law (5.28), $\tilde{w}_f(k+1) - \tilde{w}_f(k) = -\eta(k)\sigma_f[z(k)]e_i^T(k)$ and $\tilde{w}_g(k+1) - \tilde{w}_g(k) = -\eta(k)u(k)\sigma_g[z(k)]e_i^T(k)$, also from (5.29)

$$\Delta V(k) = [\| \tilde{w}_f(k) - \eta(k)\sigma_f[z(k)]e_i^T(k) \|^2 - \| \tilde{w}_f(k) \|^2]$$
$$+ [\| \tilde{w}_g(k) - \eta(k)u(k)\sigma_g[z(k)]e_i^T(k) \|^2 - \| \tilde{w}_g(k) \|^2] \quad (5.30)$$
$$= \eta^2(k) \| e_i(k) \|^2 [\| \sigma_f[z(k)] \|^2 + \| \sigma_g[z(k)]u(k) \|^2]$$
$$- 2\eta(k) \| e_i^T(k) \| [\| \tilde{w}_f(k)\sigma_f[z(k)] \| + \| \tilde{w}_g(k)\sigma_g[z(k)]u(k) \|].$$

Now using (5.25)

$$
\begin{aligned}
\Delta V(k) &= \eta^2(k) \parallel e_i(k) \parallel^2 [\parallel \sigma_f[z(k)] \parallel^2 + \parallel \sigma_g[z(k)]u(k) \parallel^2] \\
&\quad - 2\eta(k) \parallel e_i^T(k) \parallel [\beta e_i(k+1) - \xi_f - \xi_g u(k)] \\
&= \eta^2(k) \parallel e_i(k) \parallel^2 [\parallel \sigma_f[z(k)] \parallel^2 + \parallel \sigma_g[z(k)]u(k) \parallel^2] \\
&\quad - 2\eta(k) \parallel e_i^T(k) \parallel \beta \parallel e_i(k+1) \parallel +2\eta(k) \parallel e_i^T(k) \parallel [\xi_f + \xi_g u(k)].
\end{aligned}
\tag{5.31}
$$

Now let $\pi(k) = \parallel \sigma_f[z(k)] \parallel^2 + \parallel \sigma_g[z(k)]u(k) \parallel^2$, $\xi(k) = \xi_f + \xi_g u(k)$, where $\pi(k) \geq 0, 0 < \eta(k) < 1$ and if $\beta \parallel e_i(k+1) \parallel > \parallel e_i(k) \parallel$ then

$$
\begin{aligned}
\Delta V(k) &\leq -2\eta(k) \parallel e_i(k) \parallel^2 +\eta^2(k) \parallel e_i(k) \parallel^2 \pi(k) \\
&\quad + 2\eta(k) \parallel e_i^T(k)\xi(k) \parallel \\
\Delta V(k) &\leq -2\eta(k) \parallel e_i(k) \parallel^2 +\eta^2(k) \parallel e_i(k) \parallel^2 \pi(k) \\
&\quad + \eta(k) \parallel e_i^T(k) \parallel^2 +\eta(k) \parallel \xi(k) \parallel^2 \\
\Delta V(k) &\leq -\eta(k)[\parallel e_i(k) \parallel^2 \{1 - \eta(k)\pi(k)\}+ \parallel \xi(k) \parallel^2].
\end{aligned}
\tag{5.32}
$$

Now let us consider

$$
\eta(k) = \frac{\eta(k)}{1 + \pi(k)}, \eta(k) > 0, \pi(k) > 0.
\tag{5.33}
$$

Also, the modeling error $\xi(k)$ has the term input in it. This modeling error is considered to be bounded as follows:

$$
\parallel \xi(k) \parallel^2 \leq \parallel \bar{\xi}(k) \parallel^2 .
\tag{5.34}
$$

Now using the conditions (5.33) and (5.34), we can express (5.32) as

$$
\Delta V(k) \leq -\frac{\eta(k)}{1 + \pi(k)} \left[\parallel e_i(k) \parallel^2 \left(\frac{1 + (1 - \eta(k))\pi(k)}{1 + \pi(k)} \right) - \parallel \bar{\xi}(k) \parallel^2 \right].
\tag{5.35}
$$

Now if $\parallel e_i(k) \parallel^2 \geq \frac{\parallel \bar{\xi}(k)\parallel^2[1+\pi(k)]}{1+(1-\eta(k))\pi(k)}$ then $\Delta V(k) \leq 0$ with the condition that the dead zone satisfies $\beta \parallel e_i(k+1) \parallel > \parallel e_i(k) \parallel, 0 < \eta(k) < 1$. If β is selected too much big, then the dead zone becomes small. Hence, we can conclude that $V(k)$ is bounded. Also if $\eta(k) = 0$, then from (5.26) it is evident that the weights are not changed and hence they are bounded. Therefore, $V(k)$ is bounded.

5.4 Sliding-Mode Control

We define the control error as

$$e(k) = z^d(k) - z(k) = -z(k),$$

where $z^d(k)$ is the desired reference vector, for the vibration control, $z^d(k) = 0$. We propose a novel quasi-sliding-mode controller in (5.5) as

$$u(k) = \frac{1}{\hat{g}}\{-\hat{f} + K^T e(k) + \sigma \operatorname{sign}[s(k)]\}, \tag{5.36}$$

where $e(k) = [e(k+1-n) \cdots e(k)]^T$, $K = [k_n \cdots k_1]^T \in R^{n \times 1}$ which is selected such that the polynomial $\lambda^n + \sqrt{2}k_1\lambda^{n-1} + \cdots + 2^{\frac{n}{2}}k_n$ is stable, and $s(k)$ is switching function which is defined as

$$s(k) = e(k) + K^T e(k-1). \tag{5.37}$$

Theorem 5.2 *If the gain σ of the DSMC (5.36) satisfies*

$$\sigma \geq \frac{\beta H}{\|K\|}, \tag{5.38}$$

where H is the upper bound of the modeling error, β is the design parameter of the fuzzy model (5.16), K satisfies the polynomial

$$\lambda^n + \sqrt{2}k_1\lambda^{n-1} + \cdots + 2^{\frac{n}{2}}k_n$$

is stable, then the closed-loop system with SMC and fuzzy identifier is uniformly stable and the upper bound of the tracking error satisfies

$$\lim_{k \to \infty} \frac{1}{T} \sum_{k=1}^{T} \|e(k)\| \leq \frac{\|P\|}{\lambda_{\min}(Q)}\left(1 + \frac{\beta H}{\sigma}\right), \tag{5.39}$$

where P and Q are given in (5.45).

Proof We first prove that the switching function $s(k)$ is bounded. From (5.5), (5.16), and (5.19), the modeling error satisfies

$$\beta e_i(k+1) = \tilde{f} + \tilde{g}u(k). \tag{5.40}$$

Substitute the control (5.36) into the plant (5.5), the closed-loop system is

$$z(k+1) = \widehat{f} - \widetilde{f} + \frac{\widehat{g} - \widetilde{g}}{\widehat{g}}[-\widehat{f} + K^T e(k) + \sigma \operatorname{sign}[s(k)]]$$
$$= -\widetilde{f}(k) + K^T e(k) + \sigma \operatorname{sign}[s(k)] - \widetilde{g}(k)u(k).$$

The switching function (5.37) is

$$s(k+1) = e(k+1) + K^T \mathbf{e}(k) \tag{5.41}$$
$$= -z(k+1) + K^T \mathbf{e}(k).$$

Using (5.36),

$$e(k+1) + K^T \mathbf{e}(k) = -\sigma \operatorname{sign}[s(k)] + \widetilde{f}(k) + \widetilde{g}(k)u(k).$$

Using (5.40),
$$s(k+1) = -\sigma \operatorname{sign}[s(k)] + \beta e_i(k+1). \tag{5.42}$$

Since $|\operatorname{sign}[s(k)]| \le 1$ and $|e_i(k+1)| \le H$

$$|s(k+1)| \le \sigma + \beta H. \tag{5.43}$$

Because $\mathbf{e}(k) = [e(k-n+1)\cdots e(k)]^T$, and $e(k+1) = -K^T \mathbf{e}(k) + s(k+1)$,

$$\mathbf{e}(k+1) = A\mathbf{e}(k) + Bs(k+1), \tag{5.44}$$

where $A = \begin{bmatrix} 0 & 1 & 0 & \cdots & 0 \\ 0 & 0 & 1 & \cdots & 0 \\ \vdots & & \ddots & & \vdots \\ 0 & \cdots & \cdots & 0 & 1 \\ -k_n & \cdots & \cdots & \cdots & -k_1 \end{bmatrix} \in R^{nxn}$, $B = [0, \ldots 0, 1]^T \in R^{nx1}$. Because,

$$\det(sI - \alpha A) = \alpha^n k_n + \alpha^{n-1} k_{n-1}s + \cdots + \alpha k_1 s^{n-1} + s^n,$$

we select $K = [k_1 \cdots k_n]^T$ such that $\sqrt{2}A$ is stable ($\alpha = \sqrt{2}$). A stable $\sqrt{2}A$ can make the following Lyapunov equation that has positive-definite solutions for P and Q:

$$2A^T P A - P = -Q, \tag{5.45}$$

where $P = P^T > 0$, $Q = Q^T > 0$. Define the following Lyapunov function:

$$V(k) = \frac{1}{\sigma^2} \mathbf{e}^T(k) P \mathbf{e}(k), \tag{5.46}$$

where P is a solution of (5.45). Using (5.44), we calculate $\Delta V(k)$

$$
\begin{aligned}
\Delta V(k) &= \frac{1}{\sigma^2} \mathbf{e}^T(k+1) P \mathbf{e}(k+1) - \frac{1}{\sigma^2} \mathbf{e}^T(k) P \mathbf{e}(k) \\
&= \frac{1}{\sigma^2} \mathbf{e}^T(k) \left(A^T P A - P \right) \mathbf{e}(k) + \frac{2}{\sigma^2} \mathbf{e}^T(k) A^T P B s(k+1) \\
&\quad + \frac{1}{\sigma^2} B^T P B s^2(k+1).
\end{aligned}
$$

We define $K_1 = \left[1, k_1 \cdots k_n \right]^T$, from (5.41) $s(k+1) = K_1^T \mathbf{e}(k+1)$, $s(k) = K_1^T \mathbf{e}(k)$. From (5.45) and $s(k+1) = -\sigma \operatorname{sign}[s(k)] + \beta e_i(k+1)$, $\|A\| = \|B\| = 1$, and using (5.43)

$$
\Delta V(k) \le -\frac{1}{\sigma^2} \|\mathbf{e}(k)\|_Q^2 - \frac{2 \left[\sigma \|K_1\| - \beta H \right]}{\sigma^2} \|P\| \|\mathbf{e}(k)\| + \|P\| \left(1 + \frac{\beta H}{\sigma} \right)^2.
$$

From condition (5.38)

$$
\Delta V(k) \le -\frac{1}{\sigma^2} \|\mathbf{e}(k)\|_Q^2 + \|P\| \left(1 + \frac{\beta H}{\sigma} \right)^2.
$$

From [21] we known $V(k)$ is bounded, so $e(k)$ is bounded. Summarizing from 1 to T and using that $V(T) > 0$ and that $V(1)$ is a constant:

$$
V(T) - V(1) \le \sum_{k=1}^{T} -\lambda_{\min}(Q) \, \mathbf{e}^T(k) \mathbf{e}(k) + \|P\| \left(1 + \frac{\beta H}{\sigma} \right)^2
$$
$$
\lim_{T \to \infty} \frac{1}{T} \sum_{k=1}^{T} \lambda_{\min}(Q) \, \mathbf{e}^T(k) \mathbf{e}(k) \le \lim_{T \to \infty} \frac{1}{T} \left(V(1) + \|P\| \left(1 + \frac{\beta H}{\sigma} \right)^2 \right),
$$

thus (5.39) is satisfied. By the definition $\mathbf{e}(k) = [e(k-n+1) \cdots e(k)]^T$, we have that the tracking error $\mathbf{e}(k)$ is bounded.

5.5 Experimental Results

The proposed FDSMC, DSMC, and discrete PID controllers are compared. For PID controller, gains selected are as follows:

$$
K_{px} = 1800, \; K_{py} = 2000, \; K_{p\theta} = 2200, \; K_{dx} = 160
$$
$$
K_{dy} = 220, \; K_{d\theta} = 300, \; K_{ix} = 2000, \; K_{iy} = 2300, \; K_{i\theta} = 3500.
$$

All these controllers are designed to work within the range of AMD and TA. The value of σ are chosen to be 3 for the AMD and $\sigma = 0.17$ for the TA. The conditions for selecting the values of σ are from the viewpoint of Theorem 5.2. These parameters are selected in such a way that satisfactory chattering and vibration attenuation are achieved. The performance validation of these controllers is implemented by the vibration control with respect to the seismic execution on the prototype. The value of $\eta(k)$ is chosen to be 0.9. The position and velocity inputs related to the fuzzy systems are normalized in such a manner that $z(k) \in [-1, 1]$. Number of experiments carried out reveals that six rules for $\hat{f}$ and four rules for $\hat{g}$ are sufficient to sustain minimal regulation errors. The Gaussian membership function is utilized for this operation. Three membership functions are used to extract the linguistic variables from the floor position and velocity. As AMD and TA are placed on the second floor, so the position and velocity data from the second floor are utilized. The vibration of the shake table uses the Northridge earthquake signal.

We compare our controllers in three cases: (1) without any active control (No control); (2) with the TA; and (3) with both the AMD and TA (AMD+TA). Figures 5.1, 5.2, 5.3, 5.4, 5.5, 5.6, 5.7, 5.8, and 5.9 display the action of the PID control, DSMC, and FDSMC to curb the vibration along X-, Y-, and θ-directions. The control signals of DSMC and FDSMC are displayed in Fig. 5.10 and Fig. 5.11, respectively. For

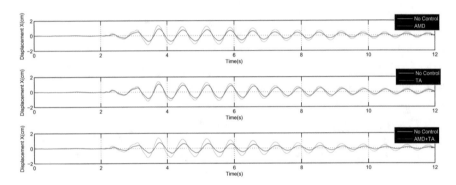

Fig. 5.1 PID control of the second floor in the X-direction

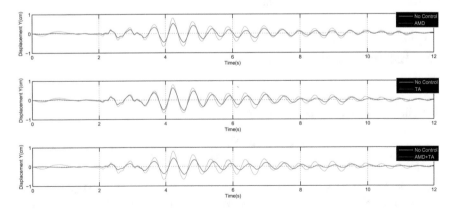

Fig. 5.2 PID control of the second floor in the Y-direction

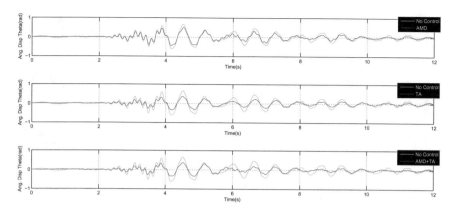

Fig. 5.3 PID control of the second floor in the θ-direction

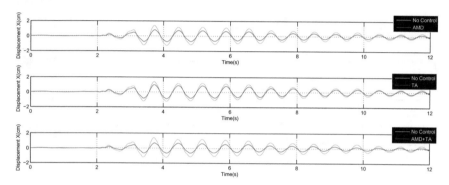

Fig. 5.4 DSMC of the second floor in the X-direction

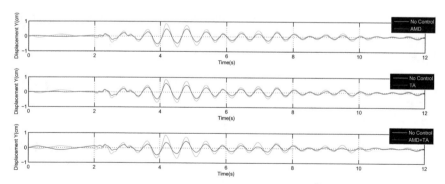

Fig. 5.5 DSMC of the second floor in the Y-direction

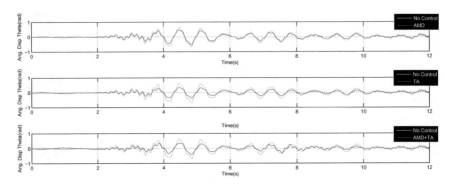

Fig. 5.6 DSMC of the second floor in the θ-direction

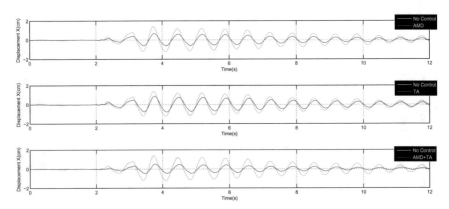

Fig. 5.7 FDSMC of the second floor in the X-direction

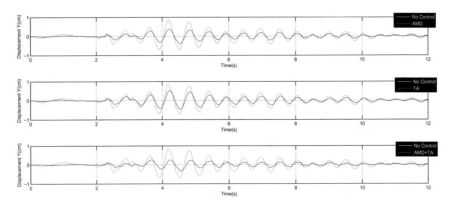

Fig. 5.8 FDSMC of the second floor in the Y-direction

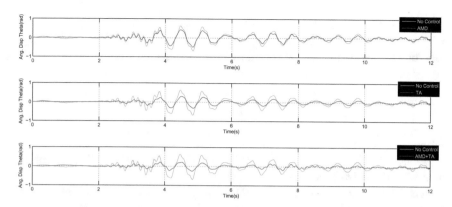

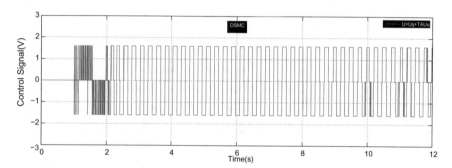

Fig. 5.9 FDSMC of the second floor in the θ-direction

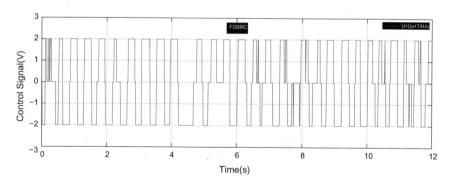

Fig. 5.10 Control signal of DSMC

Fig. 5.11 Control signal of FDSMC

clarity of the results, the vibration responses are displayed for the period of 4–10 s, whereas the control signals are scaled from the time period of 4–8 s. The average vibration displacements are calculated by the mean squared error as

$$MSE = \frac{1}{N} \sum_{k=1}^{N} x(k)^2,$$

where $x(k)$ is the displacement of the floor, and N is the total data number.

Tables 5.1, 5.2, 5.3, and 5.4 represent the quantitative analysis of vibration control along X-, Y-, and θ-directions. Here $\downarrow$ sign indicates decrease.

It is observed from the results that PID controller has slower response time in comparison to the FDSMC. We can see that all PID, DSMC, and FDSMC controllers give efficient performance with AMD and TA, because both AMD and TA act simul-

Table 5.1 Average vibration displacement by AMD+TA

Direction	PID control	% ↓ error	DSMC	% ↓ error	FDSMC	% ↓ error	No control
X	0.2216	70.5	0.1854	75.3	0.1545	79.4	0.7514
Y	0.0648	50.9	0.0436	66.9	0.0388	70.6	0.1320
θ	0.0588	49.9	0.0402	65.7	0.0312	73.4	0.1174

Table 5.2 Average vibration displacement with PID control

Direction	With AMD	% ↓ error	With TA	% ↓ error	No control
X	0.2937	60.9	0.4501	40.9	0.7514
Y	0.0760	44.6	0.0838	36.5	0.1320
θ	0.0713	39.2	0.0593	49.4	0.1174

Table 5.3 Average vibration displacement with DSMC

Direction	With AMD	% ↓ error	With TA	% ↓ error	No control
X	0.2388	68.2	0.4118	45.2	0.7514
Y	0.0577	56.2	0.0779	40.9	0.1320
θ	0.0637	45.7	0.0429	63.4	0.1174

Table 5.4 Average vibration displacement with FDSMC

Direction	With AMD	% ↓ error	With TA	% ↓ error	No control
X	0.1995	73.4	0.3847	48.8	0.7514
Y	0.0511	61.2	0.0715	45.8	0.1320
θ	0.0599	48.9	0.0399	66.1	0.1174

taneously. The vibration attenuations along X- and Y-directions are much better than θ-direction, because the active vibration control is achieved by the position of the actuator and the torque direction of the actuator. FDSMC controller is better than both PID and DSMC controllers in the vibration attenuation in all three directions. The active control of structures is improved by adding the fuzzy compensation.

5.6 Conclusions

In this chapter, the equation of motion related to the controlled building structures is converted into the form of discrete-time system. The DSMC along with fuzzy control techniques are blended to achieve superior vibration control considering bidirectional seismic forces. In the control signal (5.36), the time-varying gain helps in reducing the chattering better in comparison to the standard DSMC. A two-floor structure associated with one AMD and one TA for active vibration control is proposed. The stability of the proposed controller has been established using the Lyapunov stability theory. The theoretical analysis shows the effectivity of the proposed controllers. The experimental results show that PID, DSMC, and FDSMC controllers work well with AMD and TA. The FDSMC controller in combination with both AMD and TA is considered to be the most efficient in mitigation of vibration along X-, Y-, and θ-directions.

References

1. G.W. Housner, L.A. Bergman, T.K. Caughey, A.G. Chassiakos, R.O. Claus, S.F. Masri, R.E. Skeleton, T.T. Soong, B.F. Spencer Jr., J.T.P. Yao, Structural control: past, present and future. J. Eng. Mech. **123**(9), 897–971 (1997)
2. S. Paul, W. Yu, X. Li, Recent advances in bidirectional modeling and structural control. Shock Vibr. **2016**, 17 (2016)
3. V.I. Utkin, *Sliding Modes in Control and Optimization* (Springer, Berlin, 1992)
4. T.H. Nguyen, N.M. Kwok, Q.P. Ha, J. Li, B. Samali, Adaptive sliding mode control for civil structures using magnetorheological dampers, in *International Symposium on Automation and Robotics in Construction,* (2006)
5. K. Iwamoto, K. Yuji, N. Kenzo, K. Tanida, I. Iwasaki, Output feedback sliding mode control for bending and torsional vibration control of 6-story flexible structure. JSME Int. J. Ser. C **45**(1), 150–158 (2002)
6. M. Soleymani, A.H. Abolmasoumi, H. Bahrami, A. Khalatbari-S, E. Khoshbin, S. Sayahi, Modified sliding mode control of a seismic active mass damper system considering model uncertainties and input time delay. J. Vibr. Control. Accessed 6 July 2016
7. G. Maria, N. Selvaganesan, B. Ajith Kumar, S. Kapoor, Dynamic analysis and sliding mode vibration control for a two storeyed flexible building structure, in *International Conference on Control Communication & Computing India (ICCC),*Trivandrum, India (2016)
8. J. Dai, Fuzzy logic based control and simulation analysis for nonlinear structure, in *Second International Conference on Computer Modeling and Simulation,* Sanya, Hainan, China (2010)
9. S.Z. Sarpturk, Y. Istefanopolos, O. Kaynak, On the stability of discrete-time sliding mode control systems. IEEE Trans. Autom. Control **32**, 930–932 (1987)

10. A. Bartoszewicz, Discrete-time quasi-sliding mode control strategies. IEEE Trans. Ind. Electr. **45**, 633–637 (1998)
11. S. Janardhanan, B. Bandyopadhyay, Multirate feedback based quasi-sliding mode control of discrete-time systems. IEEE Trans. Autom. Control **52**, 499–503 (2007)
12. W. Gao, Y. Wang, A. Homaifa, Discrete-time variable structure control systems. IEEE Trans. Ind. Electron. **42**, 117–122 (1995)
13. X. Lu, B. Zhao, Discrete-time variable structure control of seismically excited building structures. Earthq. Eng. Struct. Dyn. **30**, 853–863 (2001)
14. G. Cai, J. Huang, Discrete-time variable structure control method for seismic-excited building structures with time delay in control. Earthq. Eng. Struct. Dyn. **31**, 1347–1359 (2002)
15. Z. Li, Z. Gu, Z. Deng, Discrete-time fuzzy variable structure control for buildings with delay time in control, in *8th World Congress on Intelligent Control and Automation (WCICA)*, Jinan, China, (2010)
16. Y. Fang, J. Fei, T. Hu, Adaptive backstepping fuzzy sliding mode vibration control of flexible structure. J. Low Freq. Noise, Vibr. Act. Control **37**(4), 1079–1096 (2018)
17. Z. Li, Z. Den, Improving existing "reaching law" for better discrete control of seismically-excited building structures. Front. Archit. Civ. Eng. **3**(2), 111–116 (2009)
18. J. Wang et al., Adaptive hyperbolic tangent sliding-mode control for building structural vibration systems for uncertain earthquakes. IEEE Access **6**, 74728–74736 (2018)
19. L.X. Wang, *Adaptive Fuzzy Systems and Control* (Prentice-Hall, Upper Saddle River, NJ, 1994)
20. B. Bernhard, C.J. Mulvey, A constructive proof of the Stone-Weierstrass theorem. J. Pure App. Algebra **116**(1–3), 25–40 (1997)
21. I.A. Ioannou, J. Sun, *Robust Adaptive Control* (Prentice-Hall Inc, Upper Saddle River, NJ, 1996)

Chapter 6
Bidirectional Active Control
with Vertical Effect

6.1 Introduction

During the last years, protection of civil structures subjected to different kinds of dynamic loadings is a topic of great importance and interest within the areas of active control and structural safety. To protect building structures from dynamic loadings, structural control is utilized to attenuate the undesirable vibrations to a comfortable level via actuators and dampers. Actually, the area of the structural control is classified into passive control and active control. Although there are many innovative practical implementations for passive devices, active devices are widely implemented since these are able to work in a wide range of vibration modes (a perfect choice for MDOF structures) [1, 2]. Moreover, when increasing the mass ratio of a passive actuator is not an option, an active actuator becomes an attractive alternative.

Earthquake damage reports have shown that seismic-induced building damage is often related with the torsional effect. Current investigation [3] has shown that the asymmetric feature of building structures induces a simultaneous lateral and torsional vibration known as torsion coupling when under translational excitation. Actually, it has been shown that flexible elements in asymmetric-plan building having large eccentricities in both directions may be more severely damaged due to the seismic force [4]. Moreover, when a flexible structure is subjected to lateral seismic forces, the vertical loads contribute to the lateral displacements lead to the amplification of displacements, shears, and moments throughout the entire structure. Such amplification is caused by what is known as P-Δ effect. Indeed, it is a second-order effect that could affect significantly the structural responses, and therefore it has significant implication in structural design [5].

Factors that rule the intensity of P-Δ effect are the magnitude of gravity loads and the lateral displacements induced in the structure by earthquake forces. Most building structures are designed to experience inelastic deformation during earthquakes. However, if the total deflection, elastic and inelastic, is extreme or if the

W. Yu and S. Paul, *Active Control of Bidirectional Structural Vibration*,
SpringerBriefs in Applied Sciences and Technology,
https://doi.org/10.1007/978-3-030-46650-3_6

vertical loads are bulky, the P-Δ effect may lead to structural instability. Because inelastic performance is often relied on to dissipate energy, the destabilization effect of gravity becomes exceptionally significant.

Several control algorithms are been developed for the active vibration control of building structures. For example, a semi-active PD control with magnetorheological (MR) dampers [6], fuzzy logic and PD controllers with ATMD against earthquakes [7], active tendon control using PID-type controllers [8], and feedback control strategies using PID-type controllers for active seismic control [9]. Nevertheless, the main drawbacks of these approaches are as follows: (a) they do not consider the lateral–torsional control mechanism and do not analyze the stability conditions of the closed-loop system; (b) the combination of active mass damper (AMD) and torsional actuator (TA) is not considered; and (c) dynamic responses and control calculations of structural systems are often simplified by ignoring the vertical component of a ground motion. This has been a widely accepted assumption in the engineering community since the vertical components are small enough to neglect. However, recent works have indicated that the vertical component has an important role in some earthquakes (specifically in near-fault regions earthquakes) [10].

Although the dominant research on structural control applications has been based on seismic analysis considering only unidirectional seismic waves, few advances on bidirectional seismic waves have been developed [11]. For example, a fuzzy discrete-time sliding-mode control (FDSMC) for bidirectional active control is developed in [12], PD/PID control for bidirectional active control in [13], and a PD/PID control for bidirectional active control with type-2 fuzzy system in [14]. This study, however, presents the following shortcomings: (1) do not consider the P-Δ effect and do not consider an exact model of the building structure. Doubtless, an exact model can determine the structural response effectively, and it can be helpful in control design; and (2) the vertical seismic input and its effects are not considered.

While there is no doubt about the advance in structural control, the nonlinear behavior of irregular buildings under tridirectional seismic inputs cannot be represented in a satisfactory way with an oversimplified bidirectional model. Moreover, several studies related to the control design have shown predilection with external models for control design. The model-based control system design, however, expresses a clear preference for internal descriptions of systems (state-space representations). For instance, the feedback linearization can be used as a nonlinear design tool and then use the well-known and powerful linear design [15]. In this way, state-space representations are easier to control, which make state-space representations even more efficient and effective.

For the reasons mentioned above and since control functions need to be implemented through digital devices, this chapter considers the problem of (i) finding an equivalent controllable linear model of an n-floor unsymmetric-plan building structure based on its nonlinear hysteretic discrete-time state-space representation; (ii) designing a stable active discrete-time controller via the classical pole assignment technique to attenuate the undesired vibrations of the controllable building structure into a comfortable level.

6.2 Tridimensional Model of Structures

The idealized n-floor building model considered in this chapter is depicted in Fig. 6.1. The ith ($i = 1, 2, ...n$) floor of this building consists of a rigid floor deck supported on massless axially lateral load resisting elements oriented along the direction of the ground motion as well as perpendicular to the ground motion. The position of the ith element with respect to the ith center of mass CM is shown in Fig. 6.1. Note that the building structure considered here will be subjected to horizontal ground motions, and therefore the resisting elements will be modeled using the Bouc–Wen hysteretic model. Additionally, note that two resisting elements along the direction of the earthquake are sufficient to model dynamic response of the building. Because the center of mass CM and center of stiffness CS of the floors are not located at the same location, the lateral forces will induce a coupled lateral–torsional motion. This effect arises from the asymmetric feature of the building structure; see Fig. 6.1. The center of stiffness CS, which is located in (e_x, e_y), defines the point where applied lateral forces will result only in translation of the deck.

Under bidirectional horizontal excitations, the motion of the n-floor building structure considered in this chapter is expressed as [9]

$$M\ddot{x}(t) + C\dot{x}(t) + F_s = F_{e_b}(t), \tag{6.1}$$

where $\ddot{x}(t)$, $\dot{x}(t)$, and $x(t)$ are the relative acceleration, velocity, and displacement of the floor masses, respectively; $x(t) = [X\ Y\ \Theta]^T \in R^{3n}$; M and $C \in R^{(3n)\times(3n)}$ are the mass and damping matrices, respectively; F_s is the structure stiffness force and $F_{e_b}(t)$ is the bidirectional external force applied to the building structure in X- and Y-directions. Note that

$$X = \begin{bmatrix} X_{1x} \\ X_{2x} \\ \vdots \\ X_{nx} \end{bmatrix}; \ Y = \begin{bmatrix} Y_{1y} \\ Y_{2y} \\ \vdots \\ Y_{ny} \end{bmatrix}; \ \Theta = \begin{bmatrix} r_1\theta_{1\theta} \\ r_2\theta_{2\theta} \\ \vdots \\ r_n\theta_{n\theta} \end{bmatrix},$$

Fig. 6.1 Idealized n-floor building (left); the ith floor of the building structure (right)

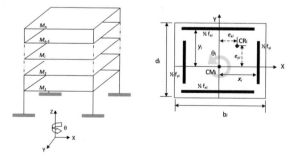

where r_i is the radius of gyration of the ith floor about a vertical axis through the center of mass of the floor. Furthermore, observe that the mass matrix is defined as $M = diag(M_x, M_y, M_\Theta)$ where $diag(\cdot)$ stands for a diagonal matrix;

$$Mx = M_y = M_\Theta = \begin{bmatrix} m_1 & 0 & 0 & 0 \\ 0 & m_2 & 0 & 0 \\ 0 & 0 & \ddots & 0 \\ 0 & 0 & 0 & m_n \end{bmatrix}$$

and

$$F_{e_b}(t) = F_{e_{XY}}(t) = \begin{bmatrix} -M_x & 0 \\ 0 & -M_y \\ 0 & 0 \end{bmatrix} \begin{bmatrix} a_{gx} \\ a_{gy} \end{bmatrix},$$

where m_i $(i = 1, 2, 3, ..., n)$ is the mass of the ith floor and a_{gx} and a_{gy} are the ground acceleration sub-vector in X- and Y-directions, respectively.

As is mentioned in , the structure stiffness force F_s can be represented as a linear or nonlinear model. Observe that if the relationship between the lateral force F_s and deformation $x(t)$ is linear then

$$F_s = Kx(t), \tag{6.2}$$

where

$$K = \begin{bmatrix} K_{xx} & 0 & -K_{x\theta} \\ 0 & K_{yy} & K_{y\theta} \\ -K_{x\theta}^T & K_{y\theta}^T & K_{\theta\theta} \end{bmatrix}.$$

Actually, the stiffness sub-matrices can be expressed as

$$K_{xx} = \begin{bmatrix} k_{x_1} + k_{x_2} & -k_{x_2} & 0 & & \\ -k_{x_2} & k_{x_2} + k_{x_3} & -k_{x_3} & & \\ 0 & -k_{x_3} & k_{x_3} + k_{x_4} & & \\ & & & \ddots & -k_{x_n} \\ & & & -k_{x_n} & k_{x_n} \end{bmatrix}$$

$$K_{yy} = \begin{bmatrix} k_{y_1} + k_{y_2} & -k_{y_2} & 0 & & \\ -k_{y_2} & k_{y_2} + k_{y_3} & -k_{y_3} & & \\ 0 & -k_{y_3} & k_{y_4} + k_{y_4} & & \\ & & & \ddots & -k_{y_n} \\ & & & -k_{y_n} & k_{y_n} \end{bmatrix}$$

$$K_{x\theta} = \begin{bmatrix} \frac{1}{r_1}(k_{x_1}e_{y_1}+k_{x_2}e_{y_2}) & -\frac{1}{r_2}k_{x_2}e_{y_2} & 0 \\ -\frac{1}{r_1}k_{x_2}e_{y_2} & \frac{1}{r_2}(k_{x_2}e_{y_2}+k_{x_3}e_{y_3}) & -\frac{1}{r_3}k_{x_3}e_{y_3} \\ 0 & -\frac{1}{r_2}k_{x_3}e_{y_3} & \ddots & & \ddots \\ & & & \ddots & & \ddots & -\frac{1}{r_n}k_{x_n}e_{y_n} \\ & & & & & -\frac{1}{r_{n-1}}k_{x_n}e_{y_n} & \frac{1}{r_n}k_{x_n}e_{y_n} \end{bmatrix}$$

$$K_{y\theta} = \begin{bmatrix} \frac{1}{r_1}(k_{y_1}e_{x_1}+k_{y_2}e_{x_2}) & -\frac{1}{r_2}k_{y_2}e_{x_2} & 0 \\ -\frac{1}{r_1}k_{y_2}e_{x_2} & \frac{1}{r_2}(k_{y_2}e_{x_2}+k_{y_3}e_{x_3}) & -\frac{1}{r_3}k_{y_3}e_{x_3} \\ 0 & -\frac{1}{r_2}k_{y_3}e_{x_3} & \ddots & & \ddots \\ & & & \ddots & & \ddots & -\frac{1}{r_n}k_{y_n}e_{x_n} \\ & & & & & -\frac{1}{r_{n-1}}k_{y_n}e_{x_n} & \frac{1}{r_n}k_{y_n}e_{x_n} \end{bmatrix}$$

and

$$K_{\theta\theta} = \begin{bmatrix} \frac{1}{r_1^2}(k_{\theta 0_1}+k_{\theta 0_2}) & -\left(\frac{1}{r_1 r_2}\right)k_{\theta 0_2} & 0 \\ -\left(\frac{1}{r_1 r_2}\right)k_{\theta 0_2} & \frac{1}{r_2^2}(k_{\theta 0_2}+k_{\theta 0_3}) & -\left(\frac{1}{r_2 r_3}\right)k_{\theta 0_3} \\ 0 & -\left(\frac{1}{r_2 r_3}\right)k_{\theta 0_3} & \ddots & & \ddots \\ & & & \ddots & & \ddots & -\left(\frac{1}{r_{n-1}r_n}\right)k_{\theta 0_n} \\ & & & & -\left(\frac{1}{r_{n-1}r_n}\right)k_{\theta 0_n} & \left(\frac{1}{r_n^2}\right)k_{\theta 0_n} \end{bmatrix},$$

where $k_{\theta 0_i} = k_{\theta_i} + k_{x_i}e_{y_i}^2 + k_{y_i}e_{x_i}^2$ for $i = 1, ..., n$.

When a building structure is excited by strong earthquakes excitations, the relationship between the lateral force F_s and the resulting deformation $x(t)$ is nonlinear. In this scenario, the stiffness component is said to be inelastic. In this way, the lateral load resisting elements under bidirectional seismic excitations can be modeled using the Bouc–Wen model [16]. In fact, the relationship between the lateral forces and displacements is

$$F_s(x(t), \dot{x}(t)) = \tilde{\alpha}_i k_i x_i(t) + (1 - \tilde{\alpha}_i)k_i f_i(t), i = 1, ..., n \tag{6.3}$$

and

$$\dot{f}_i(t) = \frac{\dot{x}_i(t) - \tilde{v}_i\left[\tilde{\beta}_i + \tilde{\gamma}_i sgn(\dot{x}_i(t)f_i(t))\right]f_i(t)\,|\dot{x}_i(t)|\,|f_i(t)|^{\tilde{n}_i-1}}{\tilde{\eta}_i}, i = 1, ..., n$$

in which $f_i(t)$ is the ith nonlinear time dependent restoring force, $\tilde{\beta}_i$, $\tilde{\gamma}_i$, $\tilde{v}_i$, $\tilde{\eta}_i$ and $\tilde{n}_i$ are parameters that control the hysteresis shape and the degradation of the system. Moreover, the variables $\tilde{\alpha}_i$, $\tilde{\eta}_i$ and k_i control the initial tangent stiffness.

The matrix C for this chapter is defined as a Rayleigh damping coefficient matrix [17]. That is, using the Rayleigh damping $C = a_0 M + a_1 K$. Indeed, the damping coefficient matrix is defined as

$$
C = \begin{bmatrix} C_{xx} & 0 & -C_{x\theta} \\ 0 & C_{yy} & C_{y\theta} \\ -C_{x\theta}^T & C_{y\theta}^T & C_{\theta\theta} \end{bmatrix}.
$$

To consider the vertical effects in the nonlinear model described in equation (6.1), let us think of the system as an MDOF oscillator [18]. When this oscillator is under vertical and translational seismic input excitations, the resultant reacting forces of the corresponding fixed-base model can be described as contributions of the induced effects of the vertical seismic excitation, the vertical oscillator response, and the gravity loads ($P - \Delta$ effect).

Here, $\phi = X_{ix}/h_i$ for $i = 1, \ldots, n$ for small values of ϕ ($\sin \phi \approx \phi$). Hence, the corresponding equation of motion (6.1) in X- and Y-directions can be rewritten as

$$
M\ddot{x}(t) + C\dot{x}(t) + F_s(x(t), \dot{x}(t)) = F_{e_b} + F_v, \tag{6.4}
$$

where

$$
F_v = \begin{pmatrix} \frac{M_x}{h} \left(-\ddot{v}_g(t) - \ddot{v}(t) + g \right) X(t) \\ \frac{M_y}{h} \left(-\ddot{v}_g(t) - \ddot{v}(t) + g \right) Y(t) \\ 0 \end{pmatrix}
$$

denotes the contributions of the induced vertical effects; $\frac{M_x}{h} = \frac{M_y}{h} = diag(\frac{m_1}{h_1}, \ldots, \frac{m_n}{h_n})$; $\ddot{v}_g(t) = a_{gz}$ is the ground acceleration in Z-direction; $\ddot{v}(t)$ is the acceleration of the oscillator; and g is the standard acceleration due to gravity.

Assuming that the vertical dynamic equation has a noticeable stiffness, the vertical dynamic equation can be assumed to behave in the linear range.

Based on the above assumption, the coupled vertical equation of motion for system (6.4) is given by

$$
M_v\ddot{v}(t) + C_v\dot{v}(t) + K_v v(t) = F_{ez}(t), \tag{6.5}
$$

where $\ddot{v}(t)$, $\dot{v}(t)$, and $v(t)$ are the acceleration, velocity, and displacement, respectively, in the vertical direction Z; $v(t) = [Z_X \ Z_Y]^T \in R^{2n}$; $M_v \in R^{2n \times 2n}$, $C_v \in R^{2n \times 2n}$, and $K_v \in R^{2n \times 2n}$ are the mass, damping (computed using the Rayleigh method), and stiffness matrices, respectively, in the vertical direction and F_{ez} denotes the vertical seismic excitation force applied to the structure. Furthermore, $M_v = diag(M_x, M_y)$, $C_v = diag(C_{xx}, C_{yy})$, and $K_v = diag(K_{xx}, K_{yy})$. Finally, notice that

$$
Z_X = \begin{bmatrix} Z_{1x} \\ Z_{2x} \\ \vdots \\ Z_{nx} \end{bmatrix}; Z_Y = \begin{bmatrix} Z_{1y} \\ Z_{2y} \\ \vdots \\ Z_{ny} \end{bmatrix}
$$

and

$$F_{ez}(t) = \begin{bmatrix} -M_x & 0 \\ 0 & -M_y \\ 0 & 0 \end{bmatrix} [a_{gz}],$$

where a_{gz} denotes the ground acceleration in the Z-direction.

6.3 Discrete-Time Model

According to the following approximations

$$\dot{x}(t) \approx \frac{x(k+1) - x(k)}{T}$$

$$\ddot{x}(t) \approx \frac{x(k+2) - 2x(k+1) + x(k)}{T^2}$$

and

$$\dot{v}(t) \approx \frac{v(k+1) - v(k)}{T}$$

$$\ddot{v}(t) \approx \frac{v(k+2) - 2v(k+1) + v(k)}{T^2},$$

where T is the sampling period, the Euler approximate discretization model of system (6.1) is

$$x(k+2) = -T^2 M^{-1} \left[C \left(\frac{x(k+1) - x(k)}{T} \right) + F_s \left(x(k), \frac{x(k+1) - x(k)}{T} \right) \right]$$
$$+ T^2 M^{-1} \left[D(u(k) - \tilde{\psi}) + F_{e_{xy}}(k) + F_v(k) \right] + 2x(k+1) - x(k) \tag{6.6}$$

and for system (6.5) is

$$v(k+2) = -T^2 M_v^{-1} C \left(\frac{v(k+1) - v(k)}{T} \right) - T^2 M_v^{-1} K_v v(k)$$
$$+ T^2 M_v^{-1} F_{e_z}(k) + 2v(k+1) - v(k), \tag{6.7}$$

where

$$F_s(x(k), \tilde{x}) = \tilde{\alpha}_i k_i x_i(k) + (1 - \tilde{\alpha}_i) k_i f_i(k), i = 1, ..., n$$

and

$$f(k+1) = f(k) + \frac{T}{\eta_i}\left[(\tilde{x})_i - \tilde{v}_i\left[\tilde{\beta}_i + \tilde{\gamma}_i sgn(\tilde{x}_i f_i(k))\right]f_i(k)|\tilde{x}_i||f_i(k)|^{\tilde{n}_i-1}\right], i = 1, ..., n.$$

Notice that $\tilde{x} = \frac{x(k+1)-x(k)}{T}$.

Moreover, let us define the following state variables $x_1(k)$ and $x_2(k)$:

$$x_1(k) = x(k)$$
$$x_2(k) = \frac{x(k+1) - x(k)}{T}.$$

Conversely, define the following state variables $v_1(k)$ and $v_2(k)$ as

$$v_1(k) = v(k)$$
$$v_2(k) = \frac{v(k+1) - v(k)}{T}.$$

Using the last equations, systems (6.6) and (6.7) can be described in the state-space representation as

$$x_1(k+1) = x_1(k) + Tx_2(k) \tag{6.8}$$
$$x_2(k+1) = \left(I - TM^{-1}C\right)x_2(k) - TM^{-1}F_s\left(x_1(k), x_2(k)\right) + TM^{-1}(D(u(k) - \tilde{\psi}) + F_{e_{xy}}(k) + F_v(k)).$$
$$y_b(k) = x_1(k)$$

and

$$v_1(k+1) = v_1(k) + Tv_2(k) \tag{6.9}$$
$$v_2(k+1) = \left(I - TM_v^{-1}C_v\right)v_2(k) - TM_v^{-1}K_v v_1(k) + TM_v^{-1}F_{e_z}(k).$$
$$y_v(k) = v_1(k),$$

where

$$F_s\left(x_1(k), x_2(k)\right) = \tilde{\alpha}_i k_i x_{1_i}(k) + (1 - \tilde{\alpha}_i)k_i f_i(k), i = 1, ..., n$$

with

$$f(k+1) = f(k) + \frac{T}{\eta_i}\left[x_{2_i}(k) - \tilde{v}_i\left[\tilde{\beta}_i + \tilde{\gamma}_i sgn(x_{2_i}(k) f_i(k))\right]f_i(k)|x_{2_i}(k)||f_i(k)|^{\tilde{n}_i-1}\right], i = 1, ..., n$$

and

$$F_v(k) = \begin{pmatrix} \frac{M_x}{h}\left(-a_{gz}(k) - v_2(k+1) + g\right)X(k) \\ \frac{M_y}{h}\left(-a_{gz}(k) - v_2(k+1) + g\right)Y(k) \\ 0 \end{pmatrix}.$$

It is not difficult to see that the smaller the sampling period T, the higher the approximation level.

6.4 Tridimensional Active Control

The structural bidirectional model with coupled vertical excitation subjected to a three-dimensional ground acceleration $(a_{g_X}, a_{g_Y}, a_{g_\theta})$, and an active control force vector $u(t)$ is described in this study as

$$M\ddot{x}(t) + C\dot{x}(t) + F_s(x(t), \dot{x}(t)) = D(u(t) - \psi) + F_{e_b}(t) + F_v(t) \quad (6.10)$$

and

$$M_v\ddot{v}(t) + C_v\dot{v}(t) + K_v v(t) = F_{e_z}(t), \quad (6.11)$$

where $u \in R^{3n}$ is the control signal which is fed to the dampers, $\psi \in R^{3n}$ is the damping and friction vector, and $D \in R^{3n \times 3n}$ denotes the location matrix of the dampers. Actually, $D = diag(d_X, d_Y, d_\Theta)$ where $d_X \in R^{n \times n}$, $d_Y \in R^{n \times n}$, $d_\Theta \in R^{n \times n}$, and

$$d_{p_{i,j}} = \begin{cases} 1, & if \ i = j = l \\ 0, & otherwise \end{cases}, \quad p = (X, Y, \Theta)$$

$\forall i, j \in \{1, \ldots, n\}; l \subseteq \{1, \ldots, n\}$. Notice that l denotes the floors (levels) on which the dampers are installed. In the case of a three-storey building ($n = 3$), if the dampers are placed on the second floor $l = \{3\}$, then $D \in R^{9 \times 9}$ with $d_{X_{3,3}} = 1$, $d_{Y_{6,6}} = 1$, and $d_{\Theta_{9,9}} = 1$. If the dampers are installed on the first, the second, and the third floors, then $l = \{1, 2, 3\}$ and $[d_X]_{3 \times 3} = [d_Y]_{3 \times 3} = [d_\Theta]_{3 \times 3} = I$. Consequently, $D = I_{9 \times 9}$.

The damper force F_{d_q}, applied by the qth damper on the structure along X- and Y-directions, is defined as

$$F_{d_q} = m_{d_q}(\ddot{x}_{d_q} + \ddot{x}_{s_q}) = u_q - \psi_q,$$

where m_{d_q} is the mass of the qth damper, $\ddot{x}_{d_q}$ is the acceleration of the qth damper, $\ddot{x}_{s_q}$ is the acceleration of the sth floor on which the damper is installed, u_q is the control signal to the qth damper, and

$$\psi_q = \mu_{d_q}\dot{x}_{d_q} + \varepsilon_q m_{d_q} g \tanh[\beta \dot{x}_{d_q}],$$

where μ_{d_q} and $\dot{x}_{d_q}$ are the damping coefficient and velocity of the qth damper, respectively. Moreover, notice that the second term of the last equation is the well-known Coulomb friction represented by a hyperbolic tangent where β denotes a positive constant, ε_q the friction coefficient between the qth damper and the floor on which it is attached and g is the gravity constant [19].

The second element of the control $u(t) = [u_X, u_Y, u_\Theta]^T$ is the torsional force u_Θ. The inertia moment of the qth torsional actuator is

$$J_{t_q} = m_{t_q} r_{t_q}^2,$$

where m_{t_q} is the mass of the qth disk and r_{t_q} is the radius of the qth disk. The torque τ_q generated by the qth disk is

$$\tau_q = J_{t_q}(\ddot{\theta}_{t_q} + \ddot{\theta}_{s_q}) = u_{\Theta_q} - \psi_{\Theta_q},$$

where $\ddot{\theta}_{t_q}$ is the angular acceleration of the qth torsional actuator, $\ddot{\theta}_{s_q}$ is the angular acceleration of the sth floor of the building, and

$$\psi_{\Theta_q} = \mu_{t_q}\dot{\theta}_{t_q} + F_{t_q}\tanh[\beta\dot{\theta}_{t_q}],$$

where μ_{t_q} is the qth torsional viscous friction coefficient, F_{t_q} is the qth Coulomb friction torque, and $tanh$ is the hyperbolic tangent depending on a positive constant β and $\dot{\theta}_{t_q}$ (the qth motor speed) [19].

Remark 6.1 Active mass damper (AMD) is a popular actuator that uses a movable mass of about 1% of the total building structure mass without a spring and dashpot. Indeed, the function of the AMD is to debilitate the vibrations of the building structure along the X- and Y-directions, while the torsional actuator (TA) decreases the torsional effect of the building. As shown in Fig. 6.2, the AMD is installed near the mass center of the building and the TA on the physical center of the building. Additionally, notice that ψ in the closed-loop system (6.10) is

$$\psi = \begin{bmatrix} \psi_{X_q} \\ \psi_{Y_q} \\ \psi_{\Theta_q} \end{bmatrix} = \begin{bmatrix} \mu_d \dot{x}_{d_{qX}} + \varepsilon_q m_{d_q} g \tanh[\beta \dot{x}_{d_{qX}}] \\ \mu_d \dot{x}_{d_{qY}} + \varepsilon_q m_{d_q} g \tanh[\beta \dot{x}_{d_{qY}}] \\ \mu_{t_q}\dot{\theta}_{t_q} + F_{t_q}\tanh[\beta\dot{\theta}_{t_q}] \end{bmatrix}. \tag{6.12}$$

Following the same procedure as above, let us find the corresponding discrete-time version of the active horizontal and torsional actuators. The discrete-time damper force F_{a_q}, applied by the qth damper on the structure along X- and Y-directions, can be defined as

$$f_{a_q}(k) = m_{d_q}(a_{d,q}(k+2) + a_{s,q}(k+2)) = u_q(k) - \tilde{\psi}_q,$$

where m_{d_q} is the mass of the qth damper, $a_{d,q}(k+2)$ is the acceleration of the qth damper, $a_{s,q}(k+2)$ is the acceleration of the sth floor on which the damper is installed, $u_q(k)$ is the discrete-time control signal applied to the qth damper and

$$\tilde{\psi}_q = \mu_{d_q} a_{d,q}(k+1) + \varepsilon_q m_{d_q} g \tanh\left[\beta a_{d,q}(k+1)\right],$$

where μ_{d_q} and $a_i(k+1)$ are the damping coefficient and velocity of the qth damper, respectively.

The second element of the discrete-time control $u(k) = [u_X, u_Y, u_\Theta]^T$ is the torsional force $u_\Theta(k)$. Therefore, the discrete-time torque $\tilde{\tau}_q$ generated by the qth disk is

$$\tilde{\tau}_q = J_{t_q}(\alpha_{t_q}(k+2) + \alpha_{s_q}(k+2)) = u_{\Theta_q}(k) - \tilde{\psi}_{\Theta_q},$$

where J_{t_q} is the inertia moment of the qth torsional actuator, $\alpha_{t_q}(k+2)$ is the angular acceleration of the qth torsional actuator, and $\alpha_{s_q}(k+2)$ is the angular acceleration of the sth floor of the building and

$$\tilde{\psi}_{\Theta_q} = \mu_{t_q}\alpha_{t_q}(k+1) + F_{t_q}\tanh[\beta\alpha_{t_q}(k+1)],$$

where $\alpha_{t_q}(k+1)$ is the qth motor speed.

Finally, observe that $\tilde{\psi}$ in the closed-loop system (6.6) has the following form:

$$\tilde{\psi} = \begin{bmatrix} \tilde{\psi}_{X_q} \\ \tilde{\psi}_{Y_q} \\ \tilde{\psi}_{\Theta_q} \end{bmatrix} = \begin{bmatrix} \mu_{d_q}a_{d,q}(k+1)_X + \varepsilon_q m_{d_q}g\tanh\left[\beta a_{d,q}(k+1)\right]_X \\ \mu_{d_q}a_{d,q}(k+1)_Y + \varepsilon_q m_{d_q}g\tanh\left[\beta a_{d,q}(k+1)\right]_Y \\ \mu_{t_q}\alpha_{t_q}(k+1) + F_{t_q}\tanh[\beta\alpha_{t_q}(k+1)] \end{bmatrix}. \quad (6.13)$$

6.5 Stability Analysis of Tridimensional Active Control

Let us pose the question of whether there exist a state feedback control $u(k)$ such that the coupled nonlinear system (6.8) can be transformed into a linear state equation of the form

$$x(k+1) = Gx(k) + Hv_u(k) \quad (6.14)$$
$$y(k) = Cx(k),$$

where the pair (G, H) is controllable.

For simplicity, let us assume that $D = I$. Hence, inspection of the state-space equation (6.8) shows that we can use

$$u(k) = D^{-1}F_s(x_1(k), x_2(k)) + \tilde{\psi} - D^{-1}F_{e_{xy}}(k) - D^{-1}F_v(k) + \frac{D^{-1}M}{T}v_u(k) \quad (6.15)$$

to cancel the undesired terms and the nonlinear term $F_s(x_1(k), x_2(k))$. Thus, this cancellation results in the linear system

$$x_1(k+1) = x_1(k) + Tx_2(k) \quad (6.16)$$
$$x_2(k+1) = \left(I - TM^{-1}C\right)x_2(k) + v_u(k)$$
$$y_b(k) = x_1(k).$$

Actually, this system can be expressed in a more compact form as

$$x(k+1) = Gx(k) + Hv_u(k) \tag{6.17}$$
$$y(k) = Cx(k),$$

where

$$G = \begin{bmatrix} I_{3n \times 3n} & T I_{3n \times 3n} \\ 0 & (I_{3n \times 3n} - TM^{-1}C) \end{bmatrix}, H = \begin{bmatrix} 0_{3n \times 3n} \\ \Gamma_{3n \times 3n} \end{bmatrix}; C = \begin{bmatrix} I_{3n \times 3n} \\ 0 \end{bmatrix}$$

and

$$\Gamma_{i,j} = \begin{cases} 1, & if\ i = j = d_s \\ 0, & otherwise \end{cases}$$

for all $i, j \in \{1, \dots, n\}$ and for all; $d_s \subseteq \{1, \dots, n\}$. Note that d_s denotes the floors on which the dampers are set.

The stabilization problem for the coupled nonlinear system (6.8) has been reduced to a stabilization problem for a controllable linear system. Therefore, we can proceed to design a stabilizing linear state feedback control

$$v_u(k) = -Kx(k) \tag{6.18}$$
$$= -K_1 x_1(k) - K_2 x_2(k),$$

to locate the eigenvalues of the closed-loop system (6.17) inside of the unit circle (pole assignment). Therefore, the overall state feedback control law is given by

$$u(k) = D^{-1} F_s(x_1(k), x_2(k)) + \ddot{\psi} - D^{-1} F_{exy}(k) - D^{-1} F_v(k) + \frac{D^{-1}M}{T}[-K_1 x_1(k) - K_2 x_2(k)],$$

where

$$K_1 = \begin{bmatrix} k_{1x} & 0 & 0 \\ 0 & k_{1y} & 0 \\ 0 & 0 & k_{1_\theta} \end{bmatrix}; K_2 = \begin{bmatrix} k_{2x} & 0 & 0 \\ 0 & k_{2y} & 0 \\ 0 & 0 & k_{2_\theta} \end{bmatrix}$$

and

$$k_{1_\rho} = \begin{bmatrix} k_{1_\rho}^1 & 0 & & \\ 0 & k_{1_\rho}^2 & & \\ & & \ddots & 0 \\ & & 0 & k_{1_\rho}^n \end{bmatrix}; k_{2_\rho} = \begin{bmatrix} k_{2_\rho}^1 & 0 & & \\ 0 & k_{2_\rho}^2 & & \\ & & \ddots & 0 \\ & & 0 & k_{2_\rho}^n \end{bmatrix}$$

for $\rho = (X, Y, \theta)$ and $i = 1, 2, 3, \dots, n$.

Specifically, the characteristic polynomials of system (6.17) are

$$z^2 + \left(k^1_{2_X} + T\frac{C_{1,1}}{m_{1_X}} - 2 \right) z + \left(-k^1_{2_X} - T\frac{C_{1,1}}{m_{1_X}} + 1 + Tk^1_{1_X} \right) = 0$$

$$\vdots$$

$$z^2 + \left(k^n_{2_X} + T\frac{C_{n,n}}{m_{n_X}} - 2 \right) z + \left(-k^n_{2_X} - T\frac{C_{n,n}}{m_{n_X}} + 1 + Tk^n_{1_X} \right) = 0$$

$$z^2 + \left(k^1_{2_Y} + T\frac{C_{n+1,n+1}}{m_{1_Y}} - 2 \right) z + \left(-k^1_{2_Y} - T\frac{C_{n+1,n+1}}{m_{1_Y}} + 1 + Tk^1_{1_Y} \right) = 0$$

$$\vdots$$

$$z^2 + \left(k^n_{2_y} + T\frac{C_{2n,2n}}{m_{n_y}} - 2 \right) z + \left(-k^n_{2_y} - T\frac{C_{2n,2n}}{m_{1_y}} + 1 + Tk^n_{1_y} \right) = 0$$

$$z^2 + \left(k^1_{2_\theta} + T\frac{C_{2n+1,2n+1}}{m_{1_\theta}} - 2 \right) z + \left(-k^1_{2_\theta} - T\frac{C_{2n+1,2n+1}}{m_{1_y}} + 1 + Tk^n_{1_\theta} \right) = 0$$

$$\vdots$$

$$z^2 + \left(k^n_{2_\theta} + T\frac{C_{3n,3n}}{m_{n_\theta}} - 2 \right) z \left(-k^n_{2_\theta} - T\frac{C_{3n,3n}}{m_{1_y}} + 1 + Tk^n_{1_\theta} \right) = 0,$$

where $C_{1,1}, \ldots, C_{n,n}, C_{n+1,n+1}, \ldots, C_{n+1}$ are the corresponding coefficients of the damping matrix C and T is the sampling period. Moreover, the last equation can be simplified as

$$z^2 + \alpha^1_{1_X} z + \alpha^1_{0_X} = 0$$

$$\vdots$$

$$z^2 + \alpha^n_{1_X} z + \alpha^n_{0_X} = 0$$
$$z^2 + \alpha^1_{1_Y} z + \alpha^1_{0_Y} = 0$$

$$\vdots$$

$$z^2 + \alpha^n_{1_Y} z + \alpha^n_{0_Y} = 0$$
$$z^2 + \alpha^1_{1_\theta} z + \alpha^1_{0_\theta} = 0$$

$$\vdots$$

$$z^2 + \alpha^n_{1_\theta} z + \alpha^n_{0_\theta} = 0,$$

where $\alpha^1_{1_X} = \left(k^1_{2_X} + T\frac{C_{1,1}}{m_{1_X}} - 2 \right), \ldots, \alpha^n_{0_\theta} = \left(-k^n_{2_\theta} - T\frac{C_{3n,3n}}{m_{1_y}} + 1 + Tk^n_{1_\theta} \right)$. That is,

$$P^i_\rho(z) = z^2 + \alpha^i_{1,\rho} z + \alpha^i_{0,\rho}, \tag{6.19}$$

for all $\rho = (X, Y, \theta)$ and for all $i = 1, \ldots, n$.

In this way, system (6.17) is stable if and only if $|z| \leq 1$ for all characteristic roots of each equation of system (6.19) and those characteristic roots z with $|z| = 1$ are simple (not repeated). Contrarily, if there is a repeated characteristic root z with $|z| = 1$, then the zero solution of the corresponding equation is unstable. Now, it is evident from the Schur–Cohn criterion that the necessary and sufficient condition for the asymptotic stability of system (6.17) is given by

$$\left| \alpha_{1_\rho}^i \right| < 1 + \alpha_{0_\rho}^i < 2$$

for all $\rho = (X, Y, \theta)$ and for all $i = 1, \ldots, n$.

Remark 6.2 When structural control is implemented in real applications, the maximum control force will be limited by the capacity of the actuators. In this way, if the control signal of the qth actuator exceeds its maximum value, it can be redefined by the following saturated control signal:

$$\bar{u}_q = \begin{cases} u_q, & |u_q| < u_{max} \\ u_{max} sgn(u_q), & otherwise \end{cases}$$

in which u_q, u_{max}, and $\bar{u}_q$ are the theoretical control signal, the maximum limit control signal, and the actual control signal of the qth actuator, respectively. Finally, notice that this new saturation controller becomes stable (i.e., is bounded in some sense).

6.6 Experimental Results

In order to analyze and validate the tridimensional control, a two-floor building structure is designed and constructed. This structure is mounted on a bidirectional shake table, see Fig. 3.2 in Chap. 3. It utilizes two Quanser one degree of freedom (DOF) shake tables for X- and Y-directions. The AMD and TA are placed on the second floor of the structure. The total moving masses of the AMD and TA are taken to be the 5% of the total mass of the building structure, see Fig. 6.2. The relative acceleration on the second floor is subtracted by the ground floor acceleration. Numerical integrators are used to compute the velocity and position from the accelerometer signal.

The parameters of this building are as follows: number of floors is $n = 2$; mass of each floor is $m = 2.9235$ kg; lateral stiffness of each floor in X-direction is $k_x = 1.3897 \times 10^6$ N/m and in the Y-direction is $k_y = 1.6754 \times 10^6$ N/m; vertical stiffness of each floor in the Z-direction is $k_z = 4.4506 \times 10^8$ N/m; the torsional stiffness of each floor in Θ-direction is $k_\theta = 4.4506 \times 10^8$ N/m; the radius of gyration of each floor is $r = 10.4083$ m; the lengths and eccentricities of each floor in X- and Y-directions are $b = 0.3$ m, $x = 0.4$ m, $e_x = 0.5$ m and $d = 2$ m, $e_y = 0.5$ m,

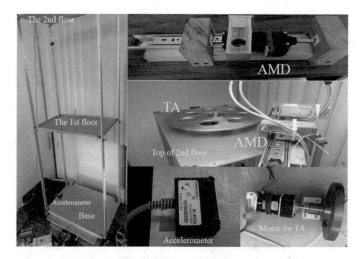

Fig. 6.2 Actuators for the bidirectional active control

$y = 0.9$ m, respectively; the heights are $h_1 = 0.7$ m for the first floor, and $h_2 = 0.2$ m for the second floor. Additionally, the hysteresis parameters of each floor are defined as $\tilde{\alpha} = 0.05$; $\tilde{\eta} = 1$; $\tilde{v} = 1$; $\tilde{\beta} = 0.5$; $\tilde{\gamma} = 0.5$; and $\tilde{n} = 2$.

The physical parameter values used for the AMD and the TA used in this numerical example are presented in Table 6.1. Observe that the total moving mass of the AMD and the TA was designed to be approximate the 0.8% (70 kg) of the total mass of the structure. In this example, the maximum control force will be bounded by the capacity of the actuators. Hence, the maximum operating voltage for the actuators was defined as $\pm 700 DC - Volts$.

Observe that for stabilization purpose, the designed stabilizing linear state feedback control (6.18) can be rewritten as

$$v_u(k) = -Kx(k)$$
$$= -K_1 x_1(k) - K_2 x_2(k) + r(k),$$

where K is the state feedback gain and $r(k)$ is the reference input. Note, however, that in the area of active vibration control of building structures, the desired reference $r(k)$ is defined to be zero.

The first 53 seconds of the North–South (NS), the East–West (EW), and the vertical (V) components of the ground acceleration record of the 1940 El Centro earthquake were used as the input excitation. In this example, the sampling period used for the Euler approximation procedure was assumed to be $T = 0.0002$ s.

Since system (6.8) is linearizable and controllable via system (6.15), the arbitrary pole assignment is possible. In this way, let us select the desired closed-loop poles at

Table 6.1 Physical parameters of the AMD and the TA

Component	Parameter	Symbol	Value
Brushed DC Motor	Coil resistance	$R[\Omega]$	0.83
	Coil inductance	$L[H]$	0.000231
	Back EMF	$V_b\left[\frac{V-s}{rad}\right]$	0.1263
	Torque constant	$K_m\left[\frac{N-m}{A}\right]$	0.1263
	Viscous friction	$B_m\left[\frac{N-m}{\frac{rad}{s}}\right]$	2.0×10^{-3}
AMD	Mass	$m_d[kg]$	25
	Gear radius	$r_m[m]$	30×10^{-3}
	Gear reduction	K_g	$1:1$
	Friction coefficient of damper	μ_d	0.020
	Friction coefficient	ε_d	0.009
	Gravitational constant	$g\left[\frac{m}{s^2}\right]$	9.80665
Torsional Actuator	Disk mass	$m_t[kg]$	20
	Disk radius	$r_t[m]$	0.9
	Inertia	$J_T[kg-m^2]$	16.20
	Friction coefficient	μ_t	0.020
	Coulomb friction	F_t	0.05
	Gear radius	$r_{m_t}[m]$	28.5×10^{-3}
	Gear reduction	K_{g_t}	$1:1$

$z = 0.9999$ and $z = 0.9001$ for each controller. That is, the corresponding characteristic polynomial of the closed-loop system is $P_\rho^3(z) = z^2 - 1.90z + 0.90001$ for all $\rho = (X, Y, \theta)$. Indeed, observe that these systems are Schur polynomials since they satisfy the condition $\left|\alpha_{1_\rho}^3\right| < 1 + \alpha_{0_\rho}^3 < 2$ for all $\rho = (X, Y, \theta)$. Finally, notice that the state feedback gain matrix is computed as

$$
K = \begin{bmatrix}
0 & 0 & 0 & 0 & 0 & 0 & 0 & 0 & 0 & 0 & 0 & 0 & 0 & 0 & 0 & 0 & 0 & 0 \\
0 & 0 & 0 & 0 & 0 & 0 & 0 & 0 & 0 & 0 & 0 & 0 & 0 & 0 & 0 & 0 & 0 & 0 \\
0 & 0 & -0.05 & 0 & 0 & 0 & 0 & 0 & 0 & 0 & -0.0995 & 0 & 0 & 0 & 0 & 0 & 0 & 0 \\
0 & 0 & 0 & 0 & 0 & 0 & 0 & 0 & 0 & 0 & 0 & 0 & 0 & 0 & 0 & 0 & 0 & 0 \\
0 & 0 & 0 & 0 & 0 & 0 & 0 & 0 & 0 & 0 & 0 & 0 & 0 & 0 & 0 & 0 & 0 & 0 \\
0 & 0 & 0 & 0 & -0.05 & 0 & 0 & 0 & 0 & 0 & 0 & 0 & -0.0994 & 0 & 0 & 0 & 0 & 0 \\
0 & 0 & 0 & 0 & 0 & 0 & 0 & 0 & 0 & 0 & 0 & 0 & 0 & 0 & 0 & 0 & 0 & 0 \\
0 & 0 & 0 & 0 & 0 & 0 & 0 & 0 & 0 & 0 & 0 & 0 & 0 & 0 & 0 & 0 & 0 & 0 \\
0 & 0 & 0 & 0 & 0 & 0 & -0.05 & 0 & 0 & 0 & 0 & 0 & 0 & 0 & 0 & 0 & -0.0988
\end{bmatrix}
$$

with

$$
\Gamma = \begin{bmatrix}
0 & 0 & 0 & 0 & 0 & 0 & 0 & 0 & 0 \\
0 & 0 & 0 & 0 & 0 & 0 & 0 & 0 & 0 \\
0 & 0 & 1 & 0 & 0 & 0 & 0 & 0 & 0 \\
0 & 0 & 0 & 0 & 0 & 0 & 0 & 0 & 0 \\
0 & 0 & 0 & 0 & 0 & 0 & 0 & 0 & 0 \\
0 & 0 & 0 & 0 & 0 & 1 & 0 & 0 & 0 \\
0 & 0 & 0 & 0 & 0 & 0 & 0 & 0 & 0 \\
0 & 0 & 0 & 0 & 0 & 0 & 0 & 0 & 0 \\
0 & 0 & 0 & 0 & 0 & 0 & 0 & 0 & 1
\end{bmatrix}.
$$

To compare our approach with the classical PD controller, this chapter also developed a discrete-time version of it using Euler approximation with $T = 0.0002$ s. Note that the proportional and derivative controller gains were set as $K_p = K_{p_x} = K_{p_y} = K_{p_\theta} = 2200$ and $K_d = K_{d_x} = K_{d_y} = K_{d_\theta} = 2500$, respectively.

The performance of each controller was evaluated by considering the average vibration displacement computed by the root mean square (RMS) as

$$
RMS = \sqrt{\frac{1}{N} \sum_{k=1}^{N} x^2(k)}, \tag{6.20}
$$

where $x(k)$ denotes the displacement of the selected floor in the desired direction and N is the total data number. The comparison of the bidirectional controllers (Pole assignment and PD control) was done as follows: (a) without any active actuator installed on the building structure (no control); (b) actuators installed on the second floor using the pole assignment controller; (c) actuators installed on the second floor using the discrete PD controller; (d) actuators installed on the second floor using the discrete pole assignment controller; and (e) actuators installed on the second floor using the discrete PD controller. In all cases, the total moving mass of the actuators was 0.8% of the total mass of the building structure.

Tables 6.2 and 6.3 show the RMS vibration displacements (RMS-VD) and the percentage reduction of the RMS-VD when the actuators are either installed on the second or the second floor of the building structure. Figure 6.3 exhibits the actions of the actuators installed on the second floor to suppress the vibrations along the X-, Y-, and $r\theta$-directions, respectively. Additionally, Fig. 6.5 exhibits the designed saturated control signals and the corresponding control force responses of the actuators via the control methodology of this chapter (Fig. 6.4).

The results suggest that both controllers can attenuate the vibrations of the building even if the actuators are installed on either the second or second floor of the building. However, the comparison results display that in both cases that the pole-placement methodology decreases the displacement vibrations much better than the PD controller. For example, Table 6.2 shows that percentage reduction of the RMS-VD using the pole assignment controller is comparatively lower than the PD controller in all directions. In the same way, one can observe from Table 6.3 that the percentage

Table 6.2 RMS-VD and percentage reduction of the RMS-VD. Actuators installed in X-, Y-, and $r\theta$-directions of the third floor

X	RMS-VD			Percentage reduction of RMS-VD	
Floor	No control (m)	Pole assignment (m)	PD control (m)	Pole assignment (%)	PD control (%)
1st	0.0035	0.0018	0.0029	48.7757	15.9936
2nd	0.0063	0.0031	0.0053	50.5268	16.2588
3rd	0.0097	0.0047	0.0081	51.8872	16.5703
Y	RMS-VD			Percentage reduction of RMS-VD	
1st	0.0024	0.0010	0.0020	57.8248	19.1428
2nd	0.0043	0.0016	0.0034	63.3008	20.2071
3rd	0.0064	0.0022	0.0051	66.6505	20.6071
$r\theta$	RMS-VD			Percentage reduction of RMS-VD	
1st	0.0119	0.0097	0.0108	18.3399	9.4169
2nd	0.0187	0.0125	0.0162	33.1667	13.5246
3rd	0.0234	0.0129	0.0198	45.0428	15.5876

Table 6.3 RMS-VD and percentage reduction of the RMS-VD. Actuators installed in X-, Y-, and $r\theta$-directions of the second floor

X	RMS-VD			Percentage reduction of RMS-VD	
Floor	No control (m)	Pole assignment (m)	PD control (m)	Pole assignment (%)	PD control (%)
1st	0.0035	0.0021	0.0032	38.7882	7.9262
2nd	0.0063	0.0039	0.0058	39.2724	8.0181
3rd	0.0097	0.0058	0.0089	39.5347	7.9591
Y	RMS-VD			Percentage reduction of RMS-VD	
1st	0.0024	0.0012	0.0022	48.5423	9.7790
2nd	0.0043	0.0021	0.0038	50.5227	10.2300
3rd	0.0064	0.0032	0.0058	49.7691	10.1973
$r\theta$	RMS-VD			Percentage reduction of RMS-VD	
1st	0.0119	0.0075	0.0110	37.2192	8.2042
2nd	0.0187	0.0092	0.0166	50.8039	10.9957
3rd	0.0234	0.0127	0.0211	45.7349	10.1171

reduction of the RMS-VD using the pole assignment procedure and the actuators installed on the second floor reduces better vibrations than the PD controller in all directions.

Remark 6.3 Although it is well known that with the use of the PD controller the regulation error can be reduced by increasing the derivative gain K_d, the cost of

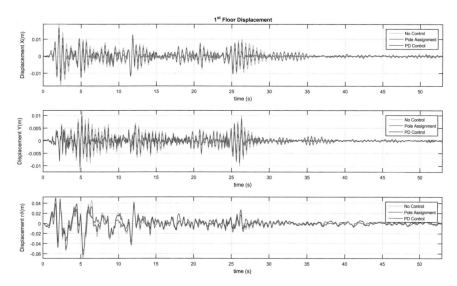

Fig. 6.3 Bidirectional and torsional responses of the first floor

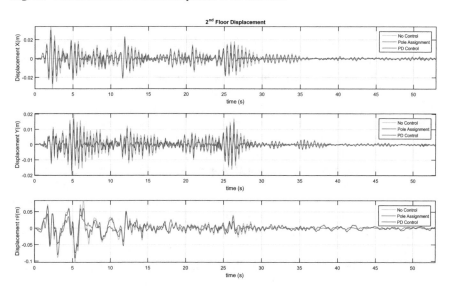

Fig. 6.4 Bidirectional and torsional responses of the second floor)

considerable values of K_d is that the transient performance becomes slow. Literally, it is only when $K_d \to \infty$ that the error converges to zero. Additionally, note that if the system compromises high-frequency noise signals, large values of K_d are not advisable.

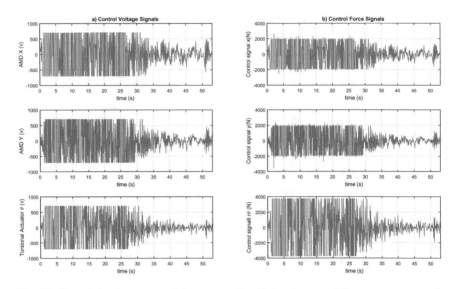

Fig. 6.5 Control signal responses of the actuators installed on the second floor

The actuators installed on the second floor of the building structure exhibits a better performance to reduce the lateral vibrations in the X- and Y-directions than those installed on the second floor of the building. However, the torsional response ($r\theta$-direction) of the actuator installed on the second floor displays a better performance to decrease the torsional vibrations than those installed on the second floor. These observations indicate that the actuators attached on the second floor attenuate mainly what is called in linear theory: the first lateral mode of vibrations (which is often dominant in the earthquake response). Meanwhile, the actuators placed on the second floor attenuate principally the torsional vibrations. In this way, one can expect to reduce either the lateral vibrations or the torsional vibrations of the building with just selecting the respective position of the actuators. In this study, the position of the actuators installed on the second floor is considered as the best position.

The numerical results show that pole assignment control design works well with the AMD and the TA. Furthermore, the results exhibit that the pole assignment controller has better performance than the classical PD in minimizing the lateral and torsional vibrations. Indeed, as can be seen from the simulation, the corresponding control design works well in the presence of nonlinear behaviors. Eventually, these results indicate that the actuators attached on the second floor are better to attenuate mainly the first lateral mode of vibrations (which is often dominant in the earthquake response) than those attached on the second floor. The results show that the second floor is the best option for placing the actuators; however, additional studies are needed to determine the optimal number and the optimal placement of the actuators on building structures.

6.7 Conclusions

This chapter studied the active control of building structures subjected to tridimensional earthquake forces. Since there are not available mathematical models for it, we develop discrete-time hysteretic nonlinear model and use the pole-placement design to control the building structure in bidirectional with vertical and torsional effects. A simple stability analysis was derived which makes this chapter even more efficient and effective to select the control gains than other controllers. The experimental results show that the effectivity of the methodology. Not any discretization technique is needed for practical implementations. The control design developed in this study is quite attractive for real implementations of active structural control. This chapter assumes that all the state variables are available. State estimators should be included in future works to estimate some unavailable variable (e.g., acceleration).

References

1. S. Elias, V. Matsagar, Research developments in vibration control of structures using passive tuned mass dampers. Ann. Rev. Control **44**, 129–156 (2017)
2. S. Thenozhi, W. Yu, Advances in modeling and vibration control of building structures. Ann. Rev. Control **37**(2), 346–364 (2013)
3. J.L. Lin, K.C. Tsai, Seismic analysis of two-way asymmetric building systems under bidirectional seismic ground motions. Earthq. Eng. Struct. Dyn. **37**(2), 305–328 (2008)
4. R.K. Goel, Seismic Response of asymmetric systems: energy-based approach. J. Struct. Eng. **123**(12) (1997)
5. E.L. Wilson, A. Habibullah, Static and dynamic analysis of multi-story buildings, including P-Delta effects. Earthq. Spectra **3**(2) (1987)
6. A. Yanik, J.P. Pinelli, H. Gutierrez, Control of a three-dimensional structure with magnetorheological dampers, in 11th International Conference on Vibration Problems, Z. Dimitrovová et al. (eds.), Lisbon, Portugal (2013)
7. R. Guclu, Sliding mode and PID control of a structural system against earthquake. Math. Comput. Model. **44**(1–2), 210–217 (2006)
8. S.M. Nigdeli, Effect of feedback on PID controlled active structures under earthquake excitations. Earthq. Struct. **6**(2), 217–235 (2014)
9. S.M. Nigdeli, M.H. Boduroglu, Active tendon control of torsionally irregular structures under near-fault ground motion excitation. Comput. Aided Civil Infrast. Eng. **28**(9), 718–736 (2013)
10. H. Kim, H. Adeli, Hybrid control of irregular steel highrise building structures under seismic excitations. Int. J. Numer. Methods Eng. **63**(12), 1757–1774 (2005)
11. S. Paul, W. Yu, X. Li, Recent advances in bidirectional modeling and structural control. Shock Vib. **2016** 17 (2016)
12. S. Paul, W. Yu, X. Li, Discrete-time sliding mode for building structure bidirectional active vibration control. Trans. Inst. Measurement Control **41**(2) 433–446 (2019)
13. S. Paul, W. Yu, A method for bidirectional active control of structures. J. Vib. Control **24**(15) 3400–3417 (2018)
14. S. Paul, W. Yu, X. Li, Bidirectional active control of structures with type-2 fuzzy PD and PID. Int. J. Syst. Sci. **49**(4), 766–782 (2018)
15. A.J. Krener, Feedback linearization of nonlinear systems. Encyclopedia Syst. Control (2013)
16. C.S. Lee, H.P. Hong, Statistics of inelastic responses of hysteretic systems under bidirectional seismic excitations. Eng. Struct. **32**(8), 2086–2074 (2010)

17. A.K. Chopra, *Dynamics of Structures*. Prentice-Hall International Series (2011)
18. A.K. Agrawal, J.N. Yang, Compensation of time-delay for control of civil engineering structures. J. Earthq. Eng. Struct. Dyn. **29**(1), 37–62 (2000)
19. C. Roldán, F.J. Campa, O. Altuzarra, et al., Automatic identification of the inertia and friction of an electromechanical actuator, in New Advances in Mechanisms, Transmissions and Applications, Volume 17 of the series Mechanisms and Machine Science, pp. 409–416 (2014)

Chapter 7
Conclusions

There has been a large amount of increased research in structural vibration control in the past few decades. A number of control algorithms and devices have been applied to the structural control applications. Linear controllers were found to be simple and effective. More advanced controllers have improved the performance and robustness. Even though this field is well developed, there is still room for further research considering lateral–torsional vibration.

In this book, an active vibration control system for building structures was developed. Three different control algorithms were developed for the structure vibration attenuation. In the first case, classical PD/PID control techniques were used to mitigate the vibration of the structure under the bidirectional forces. The stability of the controller is validated using Lyapunov candidate. In the second phase, the PD/PID control is combined with type-2 fuzzy. The PD/PID control is used to generate the control signal to attenuate the Vibration, and the type-2 fuzzy logic is used to compensate the uncertain nonlinear effects present in the system. The PD/PID gains are selected such that the system is stable in Lyapunov sense. An adaptive technique was developed for tuning the fuzzy weights to minimize the regulation error. This controller shows very good vibration attenuation capability. However, its design needs some level of system knowledge. As a result, another controller has been proposed, which can work with a parametrically uncertain system. Here, the popular sliding-mode controller has been used. So a novel fuzzy discrete sliding-mode controller (FDSMC) is proposed in order to attenuate structural vibration along all three components under the grip of bidirectional earthquake forces. The analysis is based on the lateral–torsional vibration under the bidirectional waves. The proposed FDSMC also facilities in reducing chattering due to its time-varying gain. We prove that the closed-loop system with sliding-mode control (SMC) and fuzzy identifier is uniformly stable by utilizing Lyapunov stability theorem. The proposed algorithms were experimentally verified in a lab prototype. Also, the controllers, especially the

W. Yu and S. Paul, *Active Control of Bidirectional Structural Vibration*, SpringerBriefs in Applied Sciences and Technology, https://doi.org/10.1007/978-3-030-46650-3_7

FDSMC, can function with nonlinear and uncertain systems like the real building. In the final chapter, an active control methodology is proposed for the vibration control of structures subjected to tridimensional earthquake forces.

From the experimental analysis, it has been observed that type-2 fuzzy PID controller outperformed all other controllers. But the computational cost of the controller was big. PD/PID controller also successfully attenuated the building vibration. So the PD/PID controller is highly recommended due to its simple nature. The nature of movement of sliding-mode controller is similar to that of structural movement. So this type of controller can be effectively used for vibration mitigation which is evident from the experimental results. So from computational cost and performance point of view, the FDSMC is considered to be the most reliable one. The development of novel torsional actuator (TA) offers the superior mitigation of vibration along θ component which a significant contribution in the area of torsional vibration mitigation. It is observed that both active mass damper (AMD) and TA in combination work effectively and offer efficient vibration control.

Printed in the United States
By Bookmasters